Studies in Systems, Decision and Control

Volume 137

Series editor

Janusz Kacprzyk, Systems Research Institute, Polish Academy of Sciences,
Warsaw, Poland
e-mail: kacprzyk@ibspan.waw.pl

The series "Studies in Systems, Decision and Control" (SSDC) covers both new developments and advances, as well as the state of the art, in the various areas of broadly perceived systems, decision making and control- quickly, up to date and with a high quality. The intent is to cover the theory, applications, and perspectives on the state of the art and future developments relevant to systems, decision making, control, complex processes and related areas, as embedded in the fields of engineering, computer science, physics, economics, social and life sciences, as well as the paradigms and methodologies behind them. The series contains monographs, textbooks, lecture notes and edited volumes in systems, decision making and control spanning the areas of Cyber-Physical Systems, Autonomous Systems, Sensor Networks, Control Systems, Energy Systems, Automotive Systems, Biological Systems, Vehicular Networking and Connected Vehicles, Aerospace Systems, Automation, Manufacturing, Smart Grids, Nonlinear Systems, Power Systems, Robotics, Social Systems, Economic Systems and other. Of particular value to both the contributors and the readership are the short publication timeframe and the world-wide distribution and exposure which enable both a wide and rapid dissemination of research output.

More information about this series at http://www.springer.com/series/13304

Andrés Ovalle · Ahmad Hably
Seddik Bacha

Grid Optimal Integration of Electric Vehicles: Examples with Matlab Implementation

 Springer

Andrés Ovalle
Grenoble INP, G2Elab
Université de Grenoble Alpes, CNRS
Grenoble
France

Seddik Bacha
Grenoble INP, G2Elab
Université de Grenoble Alpes, CNRS
Grenoble
France

Ahmad Hably
Grenoble INP, Gipsa-lab
Université de Grenoble Alpes, CNRS
Saint Martin d'Hères
France

ISSN 2198-4182   ISSN 2198-4190   (electronic)
Studies in Systems, Decision and Control
ISBN 978-3-319-89238-2   ISBN 978-3-319-73177-3   (eBook)
https://doi.org/10.1007/978-3-319-73177-3

Printed on acid-free paper

This Springer imprint is published by Springer Nature
The registered company is Springer International Publishing AG
The registered company address is: Gewerbestrasse 11, 6330 Cham, Switzerland

To our families

Foreword

This book is a compilation of recent research results on distributed optimization algorithms for the integral load management of plug-in electric vehicle (PEV) fleets and their potential services to the electricity system. The proposed methodologies optimally manage PEV fleets charge and discharge schedules by applying classical optimization, game theory, and evolutionary game theory techniques. It is intended to be used in graduate optimization and energy management courses.

After a short overview of the context and the state-of-the-art in Chap. 1, a classical centralized linear algorithm is presented in Chap. 2. In Chap. 3, a decentralized optimization approach using dynamic programming (DP) algorithms and a potential game framework is developed to optimally manage PEV charging schedule. Chapters 2 and 3 can be read independently. In Chaps. 4 and 5 further improvements have been introduced for decentralized load management approaches by applying two evolutionary game theory techniques: the Mixed Strategist Dynamics and the Escort Dynamics. These techniques lie in the intersection between population dynamics and game theory. They can be employed to represent the evolution of the distribution of a population of a given species over multiple territories, depending on the resources these territories offer. In these approaches, the energy consumed by a PEV, and its reactive power, are quantities represented by individual populations. Then, we have multiple populations interacting (multiple PEVs interacting) over a given set of territories. In these models, territories represent the three phases of the system at different instants ofthe day. Applying the mathematical representation of these techniques, authors proposed decentralized optimization algorithms for PEVs to provide multiple ancillary services to the electricity grid. These algorithms tackle the stochastic behavior of variables like arrival and departure, initial and final state of charge, and social or economic incentives to PEV owners from utility grid managers. Without differentiation between single and three-phase PEV chargers, the proposed methodologies seek to provide services like load compensation (load shifting and peak shaving), load balancing among phases of the system, reactive power supply, task and resource sharing among PEVs, and interaction with renewable energy micro-sources and dedicated energy storage devices.

The proposed algorithms in this book are tested under multiple scenarios using real data in collaboration with the SOREA electricity distribution company in the region of Savoie, France. The main Matlab scripts are provided and commented. These algorithms can be implemented in real-time applications and can be extended to other domains where energy management is required (smart buildings, energy storage in railways systems, task sharing of micro-generators in micro-grids, etc.). The reader can test and adapt these scripts to his/her specific applications. This book will be useful for students, researchers and engineers as well,in the domains of electric vehicle grid integration in particular, and energy management ingeneral.

Montréal, QC, Canada   Kamal Al-Haddad
January 2018   M.Sc.A. Ph.D. Fellow IEEE, Fellow CAE
                                   Professor and Canada Research Senior Chair CRC-EECPE
                                          Electric Energy Conversion and Power Electronics

Acknowledgements

Authors would like to acknowledge the work of collaborators who helped us to set the bases of the results presented in this book. It is also a pleasure to thank Gustavo Ramos, Davis Montenegro, Julian Fernandez, Kamal Al-Haddad, Alain Oustaloup, Benoît Robyns, Rachid Ibtiouen, Adrian Florescu, and anonymous reviewers whose perceptive comments have helped to improve this book immeasurably. Any flaws that remain are, of course, the responsibility of the authors.

Grenoble, France   Andrés Ovalle
Saint Martin d'Héres, France   Ahmad Hably
Grenoble, France   Seddik Bacha
January 2018

Contents

General Nomenclature

PEV	Plug-in electric vehicle
PHEV	Plug-in hybrid electric vehicle
LP	Linear programming
DP	Dynamic programming
MSD	Mixed strategist dynamics
ED	Escort dynamics
BR	Best reply or best response dynamics
i	Indices for PEVs
j	Indices for PEVs
J	Total number of PEVs
K^i	Total number of time steps for the connection time of PEV i
K	Global or general total number of time steps for time horizon
x_k^i	Power consumption/injection at time step k for PEV i (energy consumption/injection rate)
$\bar{p}^i$	Nominal power of the charger for PEV i
τ	Duration of time step k
soc_k^i	Battery state of charge for PEV i at the end of time step k
$\overline{soc}^i$	Upper limit for states of charge of PEV i at all time steps
$\underline{soc}^i$	Lower limit for states of charge of PEV i at all time steps

Chapter 1
Introduction

This introduction aims at providing an overview of the context where the algorithms and methods of this book have been developed. It also offers a description of the elements that motivate a proper management of electric vehicle fleets load and their integration to electricity distribution systems. The potential impacts and benefits that PEV fleets may bring to the grid is assessed. A brief review of the state-of-the-art methodologies with this purpose is provided. Finally, the organization of the book is outlined and each chapter contribution is described.

1.1 Context

Electrical power systems are constantly evolving with the improvement and development of technologies. In classic power systems, large quantities of energy are produced in power plants located where energy resources are available. These power plants are dispatched to optimally meet technical and economic criteria, and match forecasts of energy demand. The produced energy is transported through high voltage low current transmission infrastructures to centers of energy consumption. In these consumption centers, high voltage transported energy is transformed to medium and low voltage for its distribution to the end users. Figure 1.1a illustrates this energy flow in classic power systems.

For modern power systems, several *new devices* take place in centers of power consumption. The installation of renewable energy micro-sources dispersed throughout the electricity distributions systems is encouraged by government sustainable energy policies aiming to reduce the carbon footprint. Beside micro-sources (photo-voltaic arrays, small wind turbines, diesel generators, etc.), other new elements include dedicated energy storage systems (batteries, super-capacitors, fly-wheels, etc.), fast and robust communication infrastructures, and transportation systems with or without embedded energy storage capabilities (plug-in electric and hybrid electric vehicles,

© Springer International Publishing AG 2018 1
A. Ovalle et al., *Grid Optimal Integration of Electric Vehicles: Examples with Matlab Implementation*, Studies in Systems, Decision and Control 137, https://doi.org/10.1007/978-3-319-73177-3_1

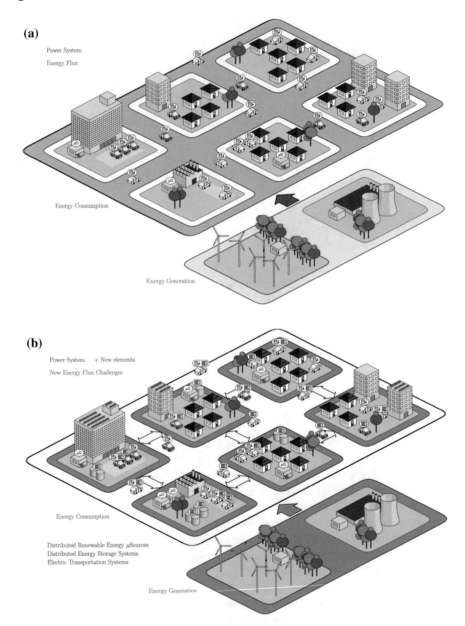

(a)

Power System

Energy Flux

Energy Consumption

Energy Generation

(b)

Power System   + New elements

New Energy Flux Challenges

Energy Consumption

Distributed Renewable Energy μSources
Distributed Energy Storage Systems
Electric Transportation Systems

Energy Generation

Fig. 1.1 Illustrative diagram of energy flow: **a** classic power systems **b** future power systems

railway electric traction systems). These devices generate, inject and consume energy within the consumption centers. Consequently, several control and modelling challenges appear on the electrical distribution infrastructure. The first challenge concerns the control of these distributed devices with a parallel improvement of efficiency at optimal costs. The second challenge concerns modelling these devices and their synergies with the power distribution system [LSH10, Mei+13, Kra+16, MM15]. This concept of *smart energy distribution system* implies that energy is moving in both time and space between different regions of the system in such an optimal way that the new technologies see their social, economic and technical values increased [Mei+13, LSH10]. To address these challenges, forecasting, smart metering, and robust communication infrastructure are the most important tools engineers and researchers have nowadays [Bas+15, PS13].

1.2 A Brief Review of the State of the Art

In a report of the International Energy Agency [Age16], 2015 was the year where the threshold of 1 million electric vehicles on the roads was exceeded (1.26 million including PEVs, PHEVs, and Fuel cell EVs). This is a curious fact since by 2013 most of the projections suggested this threshold was not going to be exceeded before 2020 [EUR14, Avi14, EP15]. Most of these vehicles are localized in the U.S., China, Japan, and European countries like Netherlands, Norway, France, and Germany, all of them members of the Electric Vehicle Initiative (EVI) group [EUR14]. In general, countries of the EVI have set a goal of 20 million sold PEVs and PHEVs by 2020. By their own, several countries have set individual goals which accumulated go up to a total of 13 million by 2020. Even if these projections account only for a market share of less than 2% of the total passenger vehicle market [Age16], after the United Nations conference on Climate Change (COP21) in 2015, nations have engaged with a decarbonisation goal of the vehicle market, and a global deployment of 100 million PEVS and PHEVs by 2030 [Age16, CC15].

In this context, fleets of plug-in electric vehicles (PEVs) and plug-in hybrid electric vehicles (PHEVs) can be considered as dynamic mobile energy storage structures serving as carriers of energy in both time and space. Their expected impacts can be critical from a local point of view: voltage sags, load unbalance, harmonics, and other power quality issues. From a global point of view, impacts expected on the grid are: frequency regulation, stability, expansion plans for upgrading infrastructure and matching supply and demand. In the most likely scenario, domestic PEV load is expected to be supplied from the residential electrical distribution system. Consequently, without proper approaches handling charging/discharging of PEVs, a high penetration rate introduces several negative impacts especially on medium to low voltage transformers [TBH14, DSB10].

1.2.1   PEV Load Management Approaches

PEV load management problem has been modeled employing several approaches and methodologies in the recent years. The most popular approaches are briefly described in this section. It is possible to identify and classify these approaches with the following descriptive tags. Each methodology falls in at least one of these categories, commonly found in the literature.

- Centralized—Decentralized: it is the most relevant label that can be applied in this context.
- Vehicle to Grid (V2G): some authors classify methodologies as V2G if PEVs are employed as dynamic movable energy storage devices, supplying ancillary services to the grid. This label is usually given if PEVs discharge their batteries to provide energy to the grid.
- Vehicle to Building (V2B): this label is given if the methodology provides services to the electrical infrastructure of a building.
- Unidirectional—Bidirectional: depending on the possibility of discharging PEV batteries, the methodology can be classified as unidirectional or bidirectional.
- On–Off—Idle: this label is given if the methodology considers only full charging or discharging rates defined by the type of charger.
- Heuristic—Rule Based: this label is given to the methodology if an approach without guarantee of reaching optimal solution is applied.
- Impact—Economic Assessment: this label is given to the methodology if it considers the analysis of the impacts on the electric infrastructure, or if it includes the economic assessment of services provided to the grid.

1.2.1.1   A Brief Review

In this section, several PEV load Management approaches will be briefly described using the category labels mentioned before. A centralized method is proposed in [CNHD10] where the objective function aims at minimizing the power losses of the grid. Several optimization approaches like Sequential Quadratic programming (SQP) and Dynamic programming (DP) are employed to find optimal schedules for the connected PEVs, considering unidirectional energy flow (discharging is not allowed). Another centralized method is proposed in [CNHD11]. Here the objective function to be minimized is linear and it represents the energy consumed for charging PEVs. All the problem constraints concerning states of charge, power boundaries and voltages are linearly formulated as well. A backward-forward sweep method is employed to compute voltages of the grid. Then, the optimization routines are applied to find optimal schedules. The procedure is repeated until the cost criteria is fulfilled.

The PEV load scheduling problem is solved under static and dynamic conditions in a centralized approach proposed in [JTG13]. Static conditions refer in this work to deterministic scenarios where all the information from PEVs (arrival and departure times, required energy, charger's power constraints, etc.) is known in advance. However, this information is not known in advance under dynamic conditions. In such a case, each time a PEV arrives, its schedule is computed knowing that schedules from previously connected PEVs have already been calculated. Under dynamic conditions, authors also propose to recalculate all the schedules each time a new PEV connects to get closer to optimality for the obtained solutions. The objective function in this work is to minimize the charging costs of PEVs while maximizing the aggregator's income.

A unidirectional centralized approach was presented in [SB12]. The approach recognizes three actors in the PEV load scheduling beside PEVs: a charging service provider (CSP), a distribution system operator (DSO), and a retailer which participates in the electricity market. Based on historical information, the CSP estimates the total amount of energy that will be consumed to charge PEVs during the day. With this estimation, the retailer optimizes its participation in the market by defining a preferred PEV load curve during the day. After receiving this preferred PEV load curve from the retailer, the CSP computes individual optimal power consumption profiles for each PEV such that the final total PEV load curve is as close as possible to the preferred load curve provided by the retailer. This scheduling process is refined in [SB10b], by adding the participation of the DSO such that grid constraints are fulfilled as well. The CSP defines individual optimal schedules by minimizing the sum of the quadratic difference and the quadratic change of the difference between the preferred PEV load curve and the actual PEV load curve (aggregating all PEVs).

Linear programming and quadratic programming for the problem of PEV load management, this time under the presence of wind power, are studied in [SB10a]. The constrained quadratic programming model is proposed to include non-linear behavior of PEV batteries in the optimization model. Assuming precise forecasts on price and wind power daily profiles the objective function aims at minimizing the cost of charging the PEV fleet. Results show that the linear approximation of the battery behavior is sufficient for the PEV load management application.

In [Ngu+14a], authors consider a centralized approach where the objective is to minimize the cost of energy for charging PEVs connected to a parking/charging station with photovoltaic modules. A binary integer programming and a linear programming approach to compute the consumption schedules of the connected PEVs are used. Binary integer programming is employed when only on/off unidirectional power flow is considered. On the other hand, the linear programming approach is proposed for the case where PEV energy consumption can be modulated instead of only considering interruptible consumption. Rules are fixed for scenarios where the photovoltaic generation is sufficient or not for charging PEVs. On the other hand, authors of [Ngu+14a] have proposed a centralized version of [Ngu+14b]. Here the objective function aims at minimizing the peak load of the parking/charging station limiting PEVs to on/off consumption rates.

A centralized strategy for a charging station is proposed in [Sar+16b]. This strategy considers bidirectional chargers assuming *charge, discharge, and idle* states for control variables. Under these conditions, an optimization model is proposed. First, the controller optimizes the subscribed power of the charging station over a year given accurate past information and estimations. In a second step, a reference ideal load profile minimizing a cost function is given. With this reference load profile, a sequential optimization approach is used. Here each PEV is attended in a queue. The schedules of each PEVs are obtained by minimizing the distance between the reference load profile of the day and the actual load profile with PEVs. Binary Integer Linear programming tools are applied to find the optimal schedule for the PEV. Once a PEV is served, the reference profile is updated by including the scheduled profile of the recently attended PEV, and the process continues with the next PEVs in the row.

A centralized supervision system combining Fuzzy-Boolean logic rules and Genetic Algorithms is proposed in [Bou+17]. The objective function is minimized by promoting local renewable energy consumption, by reducing penalties resulting from exceeding subscribed consumption and by avoiding PEV consumption during peak tariff hours. Constraints are imposed over the consumption below the subscribed power limit, renewables intermittent behavior, and deadlines for PEVs to be fully charged (5pm for working places, and 6am for residences).

Assuming PEV chargers consume energy at nominal rates from the start of their charging periods, a stochastic model for predicting the PEV load is formulated in [Zha+12]. Based on this model, a centralized optimal scheduling approach is proposed such that the variance of the total load profile (base load and PEV load) is minimized over a day horizon. The decision variable in this optimization problem is defined as the percentage of PEVs starting their charging periods at a given time step of the day, dividing the day in hour steps.

In [HVG12] a centralized method that minimizes the total cost of PEVs charging and discharging is proposed. The cost function in this work is quadratic and aims at minimizing a pricing model. PEVs arrival and departure times are uncertain and the approach requires great amounts of information to define schedules. In order to overcome this drawback, a decomposition of the global problem into several sub-problems is proposed where local central optimizers are employed. In the local objective functions, strong fluctuations in the scheduled charging rates during the charging periods are penalized. Results show that this decomposition partially reduces the limitations of the centralized approach in terms of the dependence on collecting information for executing optimization routines. However, when the number of sub-controllers is increased (more decentralized), the performance is again deteriorated.

For scenarios with low PEV penetration, in [RFK12] a comparison is given on the advantages and disadvantages of a grid model based centralized approach compared to a decentralized linear approach with constraints on voltage and PEV loads. Constraints on the decentralized approach are proposed based on off-line computed sensitivities obtained with the grid model. The proposed objectives seek to maximize the PEVs' charging rates, but leaving full final states of charge un-guaranteed.

Another interesting decentralized approach is proposed based on a communication channel analogy in [RFH14]. In this approach, each PEV divides its charging requirements in several packets of short duration at the maximal charging rate considering the owner's projected time of connection. Permissions are asked for each packet to an aggregator, which grants permits or not, depending on the availability of consumption capacity. This approach has the important advantage of fairly providing PEVs with access to the available power resources, while grid constraints are fulfilled. This work is compared with a fully centralized optimization strategy, and with a first-come-first-served approach allowing PEVs to use full rates of charge if power capacity is available. Similar to [RFK12], this kind of approaches fail to guarantee full final states of charge. Moreover, fast on/off charging rates heavily affect battery life cycles, and in high PEV penetration scenarios, performance is penalized because of the lack of flexibility of the charging rates.

Other *more* decentralized approaches are based on non-cooperative game formulations, specifically potential games [NS12]. In these scenarios, PEVs are represented as players, and the strategies they choose are represented by their power consumption profiles. Given fixed strategies from all the players, the profit of each one given its chosen strategy, and strategies from others, is defined by a function common to all the players. It has been proven that potential games have at least one equilibrium, where there is no incentive for players to deviate unilaterally from, i.e., the Nash equilibrium (NE). Furthermore, given the constraints usually imposed to feasible strategies, if the proposed potential function is strictly convex, then the NE exists and is unique [Ros65]. Under these conditions, a decentralized best reply (BR) approach where each player optimizes its response knowing fixed strategies from other players can be applied to find a NE. In [NS12], the potential function is proposed as a quadratic, variance-like cost function based on the forecast of the base load, and viable strategies consider only unidirectional power flow. Best replies from PEVs tend to allocate load on low demand hours and achieve valley filling, since uniform profiles minimize variance.

Another interesting non-cooperative game approach is presented in [MCH13] where an economic cost function is *crafted* such that the NE is reached when strategies from players achieve valley filling. This cost function is defined based on the assumption that the number of PEVs tends to infinity and even in that case their demand is not as big as the base load demand. Because of these assumptions, several drawbacks arise when the number of PEVs is not large enough to assume infinity, or when PEVs are not homogeneous in terms of time and energy needs. In these cases, the proposed function becomes non-viable to apply a BR strategy effectively. However, it was proven that this methodology is useful when fully centralized control is not a viable option.

A decentralized algorithm following a best-reply like approach similar to that introduced with [NS12] is proposed in [GTL13]. Results show that when all PEVs share similar conditions of energy requirements, charger constraints, and arrival/departure times, the methodology converge to the equilibrium in one best reply iteration for each PEV.

A local rule based method that considers only unidirectional power flow (discharging is avoided), and avoids optimization procedures is proposed in [TBH14]. This rule based algorithm defines the charging rate of the PEV at a given time as the difference between the subscribed power of the home and the power being consumed by the home appliances at that time instant. If the difference exceeds the charger's nominal power, then the charging rate is set at the limit of the charger. Additionally, every PEV charging is forbidden during peak demand hours. The impact of PEVs on the ageing rate of distribution system transformers is evaluated in [Tur+12].

In [FHB15], an assessment of the economic advantage of using PEV charging infrastructure for providing current unbalance minimization as a vehicle to grid (V2G) service is introduced. Moreover, the economic profit is included as a constraint in a centralized PEV load scheduling approach.

Even if most of the strategies mentioned above can be labelled as V2G, let us make precise reference to author labelled V2G strategies. For instance, authors of [Ota+12] propose the fleet of PEVs as a distributed spinning reserve asset, where each PEV acts according to the frequency deviation at the connection point, and centralized orders given by a *regional energy management system*. During the charging period of each PEV, a time share is dedicated only for charging while the other share is dedicated to provide the ancillary service to the grid. In [Sar+16a], a methodology for the assessment of V2G potential of PEV fleets doing daily work-home trajectories is presented. The methodology provides valuable outcomes for distribution system operators, based on the analysis of stochastic behaviors of variables like arrival and departure times, correlations between variables, and averaged quantities of the fleet charging rates, or battery capacities. It provides information under certain assumptions like fixed charging-discharging rates (modulation of charging rates is avoided), PEVs providing ancillary services once a day and being fully charged once a day, and different scenarios like at-home ancillary service providing and at-work service providing.

A load regulation and spinning reserve as the ancillary services to be provided by the PEV fleet is considered in [SES12]. From an economic point of view, scheduling is centralized by aggregators which act as interfaces between PEVs and the electricity market. PEVs charge/discharge is scheduled taking into account a cost minimization function. Then, when dispatch references for the spinning reserve are provided by the system, the aggregator splits the reference and assigns a portion to each of the PEVs. Thus, PEVs deviate their scheduled charge/discharge profiles according to the ancillary service reference provided by the aggregator.

A decentralized scheduling method, where PEVs are supposed to provide a regulation service by absorbing the uncertainties of generation and load, and smooth the power imbalance fluctuations is studied in [LLL14]. The proposed distributed algorithms are based on the gradient projection method to solve the optimization problems locally. Most of these V2G approaches must assume large number of connected PEVs, to have a substantial impact on the power system [Ota+12, SES12, LLL14, YK13].

To end this section, PEV batteries are studied in [PDK12] as dynamically configurable distributed energy storage devices acting to provide services and support to building electrical infrastructures (V2B). The potential of this dynamic energy storage systems is explored for demand side management and as emergency backup system for the building grid. With a sufficiently large fleet of PEVs and appropriate load management strategies, the aggregated battery storage system can offer an important benefit for typical building electrical infrastructures.

1.3 Book Structure

There are four chapters in this book. Each chapter contains necessary MATLAB scripts and functions. To be used as a templates for other languages, these scripts are implemented using mostly basic Matlab commands. The reader is encouraged to modify these codes and explore them for different scenarios.

Chapter 2 presents a linear centralized approach. Different from earlier contributions, the approach of this chapter is based on the linearization of the considered distribution system model under certain assumptions. The objective is to include linearized expressions for the grid voltages, as constraints of a linear optimization problem. The objective of the approach is to minimize the cost the energy employed to charge PEVs, and in parallel, their energy storage capacities are employed to provide a voltage support service to grid. The formulation of this chapter is proposed to serve as a benchmark for the identification of potential benefits and elements that can be useful for more realistic applications. Based on the conclusions of Chap. 2, the following chapters consider the separation of the scheduling problem in several distributed local optimizers corresponding to each connected PEV.

Chapter 3 describes a decentralized optimization approach where a dynamic programming (DP) algorithm is employed by each PEV to optimally manage its charging schedule. The interaction with other PEVs is based on a decentralized global optimization frame given by the application of a potential game (game theory) approach. This chapter provides a detailed explanation of a forward induction DP algorithm and its adaptation to the problem of optimal charging of a single PEV with its corresponding constraints. The extension to multiple PEVs is provided by the adaptation of a N-person non-cooperative potential game. In this game, payoffs for each player are based on a utility function aiming to reduce the distance between the total load and the average load. As a result PEV charging/discharging is scheduled such that the total load profile of the transformer is flattened (when it is feasible). The concept of *strategy* for each player in the game is carefully detailed to associate it with the dynamic programming algorithm, and with the hyper-planes where it belongs.

Chapter 4 describes an application of an evolutionary game dynamics called the Mixed Strategist Dynamics (MSD) in the decentralized PEV load scheduling problem. The proposed formulation is such that the principle of maximum entropy is applied to achieve load distributions as flat as possible given the constraints imposed by owners and chargers. Entropy measurements of the total load distribution and PEV load distributions are considered as objective functions, and a trade-off among them is defined by the PEV owners. Thus, more relevance is given to the role of PEV owners in the scheduling process, which can be economically motivated by the utility grid manager depending on how they choose to participate. While entropy maximization for the local load distributions contributes to reduce the impact on batteries' life spans, entropy maximization for the total load distribution reduces the undesirable effects over the transformer, and filters important variations on the total load. The problem is formulated such that final states of charge are assured depending on time constraints defined by the owners. An aggregator is in charge of receiving the local load distributions from PEVs, add them to the forecast of the base load and re-distribute the updated information. Moreover, the chapter introduces the concept of *fairness* in the allocation of resources among connected PEVs. Performance is evaluated using real data measured from a distribution transformer.

Chapter 5 proposes an application of an Evolutionary Game Dynamics called Escort Dynamics (ED) for the decentralized load management of Plug-in Electric Vehicles (PEV). In this chapter, PEVs work together to provide several ancillary services to the grid: load shifting, active power balancing, and partial supply of reactive power demand on each phase of the distribution transformer. Meanwhile, batteries are guaranteed to be fully charged according to constraints imposed by owners. In the proposed formulation, chargers can be either three-phase or single-phase.

Chapter 2
Centralized Approach

In this chapter, a centralized approach for plug-in electrical vehicle (PEV) fleets optimal load scheduling is described. The proposed approach considers different scenarios of PEV integration to the gird and takes into account the constraints on PEV chargers, the limits on the state of charge (SOC), and the voltage levels at the different nodes of the grid. To integrate the last constraints in this list, under certain assumptions, an approximated linear model for low voltage distribution systems is employed. The objective of the scheduling method is to minimize the energy cost employed to charge PEVs while using their energy storage capacities for voltage support to the grid. The method in this chapter serves as a benchmark for identifying the potential benefits of more elaborated approaches. Several case studies such as single and two-tariff are analyzed. The proposed centralized load scheduler is also tested under more realistic conditions using the IEEE European low voltage test feeder [Iee]. The chapter ends with different Matlab scripts allowing the reader to verify the results and test the algorithms of the chapter.

2.1 Introduction

This chapter describes a linear approach proposed for obtaining node voltages of a low voltage distribution system, based on the load profiles at each of its nodes, and the presence of PEVs. Based on this approximated linear model of the grid, a centralized PEV load scheduling approach is proposed such that the available distributed energy storage capacities of PEVs are employed to provide voltage support services to the grid [Ova+14]. This approach provides optimal charging schedules for each PEV in a centralized scheme, taking into account constraints from PEV owners perspectives. Knowing consumption forecasts, and tariff details in advance, cost function and constraints are modeled such that the optimization problem can be solved by applying linear programming techniques. The approach is tested under multiple scenarios. It is observed that, with the proper configuration of voltage constraints and tariffs,

© Springer International Publishing AG 2018
A. Ovalle et al., *Grid Optimal Integration of Electric Vehicles: Examples with Matlab Implementation*, Studies in Systems, Decision and Control 137, https://doi.org/10.1007/978-3-319-73177-3_2

the method redistributes energy consumption of PEVs providing a voltage support service to the distribution network.

The formulation of this chapter was proposed to serve as a benchmark for the identification of possible benefits and elements that could be useful for more realistic applications. Besides, it was intended to be used as a tool for compare with distributed optimization approaches like those proposed on Chaps. 3, 4, and 5.

2.2 Problem Statement

As an initial objective for this approach, it is possible to think of minimizing the cost of all the energy consumption schedules of PEVs, while constraints on charging rates (power), states of charge, and voltage support are fulfilled. First, the objective function is defined as the minimization of the sum of costs of energy consumption of each connected PEV labeled with super-index $i = \{1, 2, \ldots, J\}$, over discretized time steps $k = \{1, 2, \ldots, K^i\}$,

$$\min_{\mathbf{w}^i \in W^i, \ \mathbf{s}^i \in S^i} \sum_{i=1}^{J} \sum_{k=1}^{K^i} c_k \tau \left(w_k^i - s_k^i \right). \tag{2.1}$$

Here, parameters c_k represent the cost of energy at each time step k, τ represents the duration of time steps (in hours), J is the number of connected PEVs, and K^i is the amount of time steps of the connection time for PEV i. Variables w_k^i and s_k^i are elements of auxiliary vectors $\mathbf{w}^i$ and $\mathbf{s}^i$, whose difference,

$$x_k^i = w_k^i - s_k^i \ \forall k, \tag{2.2}$$

represents the charging/discharging rate x_k^i of each connected PEV i. Auxiliary vectors $\mathbf{w}^i$ and $\mathbf{s}^i$, are exclusively employed to represent power variables x_k^i of each PEV, at each time step, as the difference of two positive valued variables w_k^i and s_k^i. These auxiliary vectors lie within sets W^i and S^i, defined by certain boundaries, besides the positivity constraints, explained in the following subsections.

2.2.1 Constraints on the Charger

Chargers are assumed to be able to handle bidirectional power flow. Moreover, they have limited rates of consumption/injection that must be taken into account in the load scheduling problem. In this approach, for each PEV, consumption/injection rates are limited by,

$$-\overline{p}^i \leq x_k^i \leq \overline{p}^i, \ \forall k = \{1, 2, \ldots, K^i\}, \ \forall i = \{1, 2, \ldots, J\}, \tag{2.3}$$

where $\overline{p}^i$ is the nominal power of the charger of PEV i. If x_k^i is expressed as the difference of two positive variables $x_k^i = w_k^i - s_k^i$ (as in Eq. (2.3)), then constraint (2.3) can be separated in the following two constraints,

$$0 \le w_k^i \le \overline{p}^i, \quad 0 \le s_k^i \le \overline{p}^i, \quad \forall k = \{1, 2, \ldots, K^i\}, \quad \forall i = \{1, 2, \ldots, J\}, \quad (2.4)$$

which are valid for all the connection time of each PEV.

2.2.2 Constraints on Partial and Final States of Charge

For a PEV labeled i, the battery state of charge at the end of time step k, is given by,

$$soc_k^i = soc_0^i + \tau \sum_{\kappa=1}^{k}(w_\kappa^i - s_\kappa^i), \quad \forall k = \{1, 2, \ldots, K^i\}, \quad \forall i = \{1, 2, \ldots, J\}. \quad (2.5)$$

To avoid impacting the batteries states of health due to deep discharging, partial states of charge must be constrained between certain boundaries,

$$\underline{soc}^i \le soc_k^i \le \overline{soc}^i, \quad \forall k = \{1, 2, \ldots, K^i\}, \quad \forall i = \{1, 2, \ldots, J\},$$

$$\underline{soc}^i \le soc_0^i + \tau \sum_{\kappa=1}^{k}(w_\kappa^i - s_\kappa^i) \le \overline{soc}^i. \quad (2.6)$$

Here, $\overline{soc}^i$ and $\underline{soc}^i$ are the upper and lower constraints, imposed to all partial states of charge. On the other hand, the final state of charge has a more restrictive constraint. It has to be equal to a desired state of charge soc_d^i imposed by the PEV owner. This constraint is expressed as,

$$soc_0^i + \tau \sum_{k=1}^{K^i}(w_k^i - s_k^i) = soc_d^i, \quad \forall i = \{1, 2, \ldots, J\}, \quad (2.7)$$

As it can be observed, states of charge are linear functions of the decision variables of the problem, i.e. the charging/discharging rates $x_k^i = w_k^i - s_k^i$. Even if some of the constraints already mentioned are imposed over variables like the partial and final states of charge, all the constraints directly affect these power variables. In fact, these constraints are all linear constraints imposed over variables $x_k^i = w_k^i - s_k^i$.

2.2.3 Constraints on Voltage Levels

Most of the constraints mentioned before are usually employed in problems of load scheduling. However, in this chapter, a linear approximation for distribution network topologies is proposed to include the effect of instantaneous charging/discharging rates of PEVs over voltage levels. Let us consider the following constraints,

$$v_{min} \leq v_k^n \leq v_{max}, \quad \forall k = \{1, 2, \ldots, K\}, \quad \forall n = \{1, 2, \ldots, N\}. \quad (2.8)$$

Here, v_k^n represents the voltage at the node n of a grid with N nodes, at time step k. This voltage is limited to be between a lower limit v_{min} and an upper limit v_{max}. Different from the prevously employed K^i, the horizon variable K in this constraint is fixed for the centralized problem, covering all the horizons from each PEV (for instance, it can be $K = 24$ h). These voltages are non-linear functions of variables like the instantaneous active and reactive powers at each node [AWA07, Ova+15]. To find accurate instantaneous values, models of the grid elements can be employed and iterative load flow techniques can be applied [Ker01, BV00]. However, to keep the constraint set linear, in this chapter an approximation is proposed under certain assumptions.

2.3 Details on the Voltage Level Modeling Approach

Let us consider the single-line radial topology model of a residential electrical distribution system, on Fig. 2.1a. This topology has a basic feeder with an amount Ψ of nodes. Furthermore, branches may diverge from any node. For instance a branch with an amount Θ_ψ of nodes is shown, diverging from node ψ of the basic feeder. In this case, basic feeder nodes are labeled with indices $\psi = \{1, 2, \ldots, \Psi\}$, while branch nodes are labeled with pairs (ψ, θ), where $\theta = \{1, 2, \ldots, \Theta_\psi\}$. Considering the totality of nodes, let us instead consider a unified label $n = \{1, 2, \ldots, N\}$, where N is the total amount of nodes including those in the basic feeder and the branches.

Let us assume a low voltage balanced grid, with the radial topology of Fig. 2.1a. For high voltage power systems, the model of short transmission lines usually neglects the resistive portion of line impedances because the reactive portion is much larger [BV00, Gue+05, Ova+15]. On the contrary, under low voltage conditions, the resistive portion of line impedances is usually larger that the reactive portion. For the purpose of the approach of this chapter, let us assume that the resistive portion is much larger than the reactive portion ($R >> X$), so the second one can be neglected. A second assumption consists in considering only unitary power factor loads (including PEVs), eliminating the reactive portion of loads as well. Considering unitary power factor for residential load can be a strong assumption, however, it is assumed since low power factors are usually penalized with higher bill costs for electricity service customers. For a grid node labeled with index n, with active power

(a)

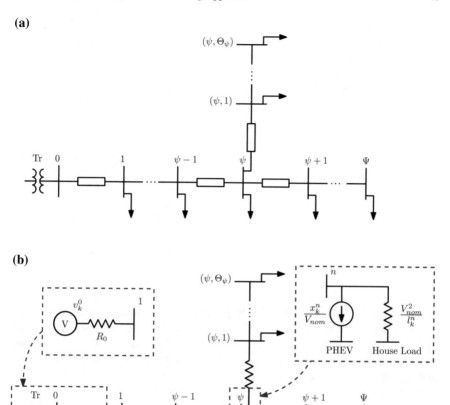

(b)

Fig. 2.1 **a** Single-line low voltage radial topology. **b** linear approximation for transformers, lines and load elements. © [2017] IEEE. Reprinted, with permission, from [Ova+14]

l_k^n at time k, the load is approximately represented with a resistance R_k^n given by,

$$R_k^n := \frac{V_{nom}^2}{l_k^n}, \tag{2.9}$$

where V_{nom} is the nominal voltage value of the grid. On the other hand, to model the effects of modulating PEVs charging/discharging rates, their batteries are represented as current sources with variable currents. Given a power reference x_k^n for a PEV connected to node n, the corresponding current value h_k^n for the model, at time k, is approximated as,

$$h_k^n := \frac{x_k^n}{V_{nom}}. \tag{2.10}$$

Fig. 2.2 Cell representing
the currents flowing through
a node in the approximated
model of the grid

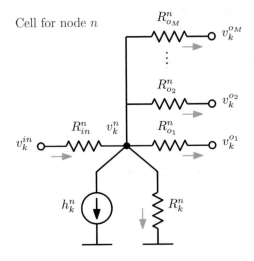

Finally, the transformer node is modeled as an fixed voltage source. After taking
into account these assumptions, the original single-line radial topology model becomes a linear circuit, as on Fig. 2.1b. The values of resistance parameters of this
linear circuit change according to the forecasts of load on each node of the grid.
Moreover, knowing the resistance parameters, the voltage at the transformer node,
and the value assigned to current sources, it is possible to find an approximation of
the voltages on each node by inverting the system matrix.

Let us decompose the radial circuit topology, on several cells defined by nodes.
Figure 2.2 shows a cell corresponding to the node n in the basic feeder. By Kirchhoff's
circuit laws, the currents on the node n define,

$$
h_k^n = v_k^{in}\left(\frac{1}{R_{in}^n}\right) - v_k^n\left(\frac{1}{R_{in}^n} + \frac{1}{R_k^n} + \frac{1}{R_{o_1}^n} + \frac{1}{R_{o_2}^n} + \cdots + \frac{1}{R_{o_M}^n}\right)
$$

$$
+ v_k^{o_1}\left(\frac{1}{R_{o_1}^n}\right) + v_k^{o_2}\left(\frac{1}{R_{o_2}^n}\right) + \cdots + v_k^{o_M}\left(\frac{1}{R_{o_M}^n}\right). \qquad (2.11)
$$

Variations on the topology of a node can be taken into account by eliminating
terms on Eq. (2.11). For instance, some options are listed as follows:

- If there is no connected PEV at the node, its expression becomes:

$$
0 = v_k^{in}\left(\frac{1}{R_{in}^n}\right) - v_k^n\left(\frac{1}{R_{in}^n} + \frac{1}{R_k^n} + \frac{1}{R_{o_1}^n} + \frac{1}{R_{o_2}^n} + \cdots + \frac{1}{R_{o_M}^n}\right)
$$

$$
+ v_k^{o_1}\left(\frac{1}{R_{o_1}^n}\right) + v_k^{o_2}\left(\frac{1}{R_{o_2}^n}\right) + \cdots + v_k^{o_M}\left(\frac{1}{R_{o_M}^n}\right).
$$

- If a node is the final node of a branch or feeder, its expression becomes:

$$h_k^n = v_k^{in} \left(\frac{1}{R_{in}^n} \right) - v_k^n \left(\frac{1}{R_{in}^n} + \frac{1}{R_k^n} \right).$$

- Finally, if a node is the first one after the transformer node, its expression becomes:

$$h_k^n - v_k^0 \left(\frac{1}{R_0} \right) = - v_k^n \left(\frac{1}{R_0} + \frac{1}{R_k^n} + \frac{1}{R_{o_1}^n} + \frac{1}{R_{o_2}^n} + \cdots + \frac{1}{R_{o_M}^n} \right)$$
$$+ v_k^{o_1} \left(\frac{1}{R_{o_1}^n} \right) + v_k^{o_2} \left(\frac{1}{R_{o_2}^n} \right) + \cdots + v_k^{o_M} \left(\frac{1}{R_{o_M}^n} \right). \quad (2.12)$$

where v_k^0 is the voltage of the transformer at time k, and R_0 is the resistance of the line connecting the transformer node and the node n.

As it can observed, there are N unknown voltage variables, and N linear equations. Organizing all the expressions for each node, as it was mentioned before, an expression with the following structure,

$$\tilde{\mathbf{A}}_k \mathbf{v}_k = \mathbf{h}_k + \left(\frac{1}{R_0} \right) \mathbf{v}_k^0 \quad (2.13)$$

can be constructed. Here, matrix $\tilde{\mathbf{A}}_k$ collects all the conductance coefficients multiplying voltage unknowns in linear Eqs. (2.11), for each node at time k. Vector $\mathbf{v}_k$ is the vector of unknown voltages organized as it was mentioned before,

$$\mathbf{v}_k = \left[v_k^1, v_k^2, \ldots, v_k^n, \ldots, v_k^N \right]^{\mathrm{T}}. \quad (2.14)$$

On the other hand, vector $\mathbf{h}_k$ gathers the current references of PEVs at each node at time step k,

$$\mathbf{h}_k = \left[h_k^1, h_k^2, \ldots, h_k^n, \ldots, h_k^N \right]^{\mathrm{T}}. \quad (2.15)$$

If there is not a connected PEV on the node n, its current entry is $h_k^n = 0$. Finally, vector $\mathbf{v}_k^0$ is a vector with the same dimensions of $\mathbf{h}_k$, whose first element is the transformer voltage at time k, and all the other elements are zero, i.e.,

$$\mathbf{v}_k^0 = \left[-v_k^0, 0, \ldots, 0 \right]^{\mathrm{T}}. \quad (2.16)$$

The sum $\mathbf{h}_k + (1/R_0) \mathbf{v}_k^0$, has its origin on expression (2.12), and it separates the contribution of the transformer (the voltage source), from the contributions of PEV batteries (current sources). The vector of currents can be expressed in terms of the power references at time k, as in Eq. (2.10). Thus, Eq. (2.13) becomes,

$$\tilde{\mathbf{A}}_k \mathbf{v}_k = \left(\frac{1}{V_{nom}}\right) \mathbf{x}_k + \left(\frac{1}{R_0}\right) \mathbf{v}_k^0, \tag{2.17}$$

where $\mathbf{x}_k = \left[x_k^1, x_k^2, \ldots, x_k^n, \ldots, x_k^N\right]^{\mathrm{T}}$, is the vector of charging/discharging rate references of PEVs. Defining $\mathbf{A}_k$ as the inverse matrix of $\tilde{\mathbf{A}}_k$, the following expression,

$$\mathbf{v}_k = \left(\frac{1}{V_{nom}}\right) \mathbf{A}_k \mathbf{x}_k + \left(\frac{1}{R_0}\right) \mathbf{A}_k \mathbf{v}_k^0, \tag{2.18}$$

provides the values of voltage unknowns for all times $k = \{1, 2, \ldots, K\}$, given information of forecasted residential load (on $\mathbf{A}_k$), and reference voltage and power values from the transformer and PEVs (on $\mathbf{v}_k^0$, and $\mathbf{x}_k$). It is important to notice that the base voltage on each node, without including PEVs, is given by the second term $\left((1/R_0)\,\mathbf{A}_k \mathbf{v}_k^0\right)$, while the first term $\left((1/V_{nom})\,\mathbf{A}_k \mathbf{x}_k\right)$ provides information on the voltage variations introduced by PEV energy consumption/injection.

With this linear approximation of voltages at every node of a low voltage grid, it is possible to know all voltages v_k^n on constraints (2.8), given by load schedules of PEVs at a time k. Thus, these voltage constraints become additional linear constraints on power of PEVs.

2.3.1 Modeling both PEV and Residential Load as a Current Sources

Instead of modeling residential loads by resistances R_k^n approximated by (2.9), they can be modeled as current sources as it was chosen for PEV load. Similar to Eq. (2.10), for the residential load the current value is approximated by,

$$g_k^n := \frac{l_k^n}{V_{nom}}, \tag{2.19}$$

where l_k^n is the active power at the node n at time k. Then, the individual cell of a node becomes the one shown on Fig. 2.3, and the expression of the node becomes,

$$h_k^n + g_k^n = v_k^{in} \left(\frac{1}{R_{in}^n}\right) - v_k^n \left(\frac{1}{R_{in}^n} + \frac{1}{R_{o_1}^n} + \frac{1}{R_{o_2}^n} + \cdots + \frac{1}{R_{o_M}^n}\right)$$
$$+ v_k^{o_1} \left(\frac{1}{R_{o_1}^n}\right) + v_k^{o_2} \left(\frac{1}{R_{o_2}^n}\right) + \cdots + v_k^{o_M} \left(\frac{1}{R_{o_M}^n}\right), \tag{2.20}$$

where all the possible variations on the topology can be considered as well. If only current sources are considered for representing both residential and PEV loads, then there are again N unknown voltage variables and N linear equations, but the matrix

Fig. 2.3 Cell representing
the currents flowing through
a node in the approximated
model of the grid.
Alternative representation of
the residential load, as a
current source

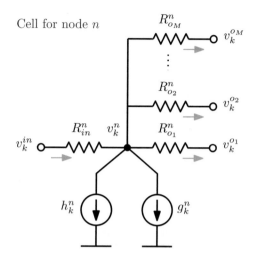

Cell for node n

of the topology becomes independent of time k. This happens because only the
resistive parts of lines are considered, and these are constant in time. In other words,
the coefficient corresponding to v_k^n on the right of Eq. (2.11) depends on time k, while
on Eq. (2.20), it is independent. As a result, organizing all the expressions for each
node provides the following expression,

$$\tilde{\mathbf{A}}\mathbf{v}_k = \mathbf{h}_k + \mathbf{g}_k + \left(\frac{1}{R_0}\right)\mathbf{v}_k^0, \tag{2.21}$$

where $\mathbf{v}_k$, $\mathbf{h}_k$, and $\mathbf{v}_k^0$, have the same constructions than before, and $\mathbf{g}_k$,

$$\mathbf{g}_k = \left[g_k^1, g_k^2, \ldots, g_k^n, \ldots, g_k^N\right]^{\mathrm{T}}. \tag{2.22}$$

is the vector of currents corresponding to residential load. After finding $\mathbf{A}$ (the inverse
of $\tilde{\mathbf{A}}$), which is valid for all time step k, and applying Eqs. (2.10) and (2.19), the
following expression,

$$\mathbf{v}_k = \left(\frac{1}{V_{nom}}\right)\mathbf{A}\mathbf{x}_k + \left(\frac{1}{V_{nom}}\right)\mathbf{A}\mathbf{l}_k + \left(\frac{1}{R_0}\right)\mathbf{A}\mathbf{v}_k^0, \tag{2.23}$$

provides the values of voltage unknowns for all times $k = \{1, 2, \ldots, K\}$, given information of forecasted residential load (on $\mathbf{l}_k$), and reference voltage and power
values from the transformer and PEVs (on $\mathbf{v}_k^0$, and $\mathbf{x}_k$). It is important to notice that
now, the base voltage on each node without including PEVs, is given by the terms
$\left((1/R_0)\,\mathbf{A}\mathbf{l}_k + (1/R_0)\,\mathbf{A}\mathbf{v}_k^0\right)$, while the first term $\left((1/V_{nom})\,\mathbf{A}\mathbf{x}_k\right)$ provides information on the voltage variations introduced by PEV energy consumption/injection.

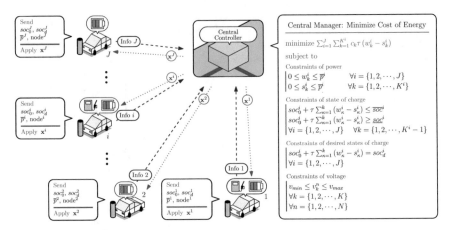

Fig. 2.4 Illustration of the interactions, and flow of information, in the proposed centralized approach for PEV load management

This alternative approximation is more flexible than the first one, especially when the amount of nodes and lines is too large. In the next section, a small example is explored to clarify the centralized approach proposed in this chapter. After this illustrative example, a more realistic scenario is considered with a large number of nodes having both residential and PEV load. To check the validity of the approximated model, the results of the proposed linear modeling approximation for the second scenario are compared against the results obtained with the well-known *OpenDSS* software for electrical distribution system modeling and analysis (from the Electric Power Research Institute—EPRI).

Before continuing with the next section, let us describe how the proposed approach works. Figure 2.4 shows an illustrative diagram for the operation of the proposed PEV load scheduling method. This scheme centralizes the optimization of schedules to minimize costs of energy consumption as it was explained before. At a given time step k, a Central Controller (CC) is in charge of gathering all the information from recently arrived PEVs. This information includes initial and desired states of charge, nominal power of the charger, node of the grid where the PEV is connected, and estimated or desired time of departure. The CC is also in charge of collecting the information of load forecast at each of the nodes of the grid where residential load is served. Once it has collected all the information, the CC runs the optimization routine and sends the consumption/injection schedules for each of the served PEVs. In the next time step k, it is likely to have newly arrived PEVs, so the procedure is repeated only with the new arrivals (those PEVs from before have their schedules already fixed).

2.4 Illustrative Example with 8-Node Grid Topology

A test grid model is shown on Fig. 2.5. This is a residential grid with eight nodes, each one serving a residence. Three PEVs are assumed to be connected to the grid at 18 h and during different charging periods (they disconnect respectively at 06 h, 04 h and 04 h30 in the next morning). They are connected on the nodes highlighted with a red dot in the grid.

The 24 h forecast of the total load profile for the transformer is shown on Fig. 2.6. The voltage at the transformer node is fixed at 230 V. Chargers are restricted to maximum charging/discharging rates of 3 kW, and batteries have nominal 20 kWh capacities. Nevertheless, only 80% of the capacity (16 kWh) is available (to reduce impact on the battery lifespans).

Parameters of lines are listed on Table 2.1. A standard low voltage cable is employed on these lines (25 mm^2 section, 0.78 Ω/km). Two cases are evaluated. On the first case, without considering PEVs connected to the grid, the loads does not cause voltage limits violations as it can be seen on Fig. 2.7. This figure shows the voltage profiles at each node, without connected PEVs. It is possible to see that each voltage profile remains between 0.9 and 1 p.u. during this period. The second case has a slight variation on certain line lengths. However, this modification is enough to cause voltage limit violations on certain nodes during peak demand hours, even without connected PEVs. For this case, the voltage profiles for each node and at each time step (without PEVs) are shown on Fig. 2.8.

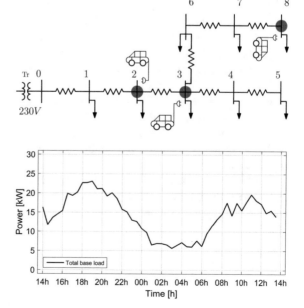

Fig. 2.5 Radial grid proposed to test the strategy. © [2017] IEEE. Reprinted, with permission, from [Ova+14]

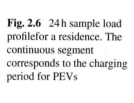

Fig. 2.6 24 h sample load profile for a residence. The continuous segment corresponds to the charging period for PEVs

Table 2.1 Grid line parameters (0.78 Ω/km)

Line		Case 1		Case 2	
From	To	Length [m]	R [Ω]	Length [m]	R [Ω]
0	1	64	0.05	64	0.05
1	2	64	0.05	64	0.05
2	3	128.2	0.10	192.3	0.15
3	4	192.3	0.15	320.5	0.25
4	5	192.3	0.15	192.3	0.15
3	6	128.2	0.10	128.2	0.10
6	7	64	0.05	64	0.05
7	8	128.2	0.10	128.2	0.10

In this study case, PEVs are assumed to have a certain initial state of charge (40%, 20% and 50% of the available capacity (16 kWh) respectively).

2.4.1 PEVs Without Charging Management—Grid Case 1

When PEVs are connected without any charging management approach, the chargers are supposed to consume at the maximum charging rate to charge the batteries as fast as possible. When batteries are fully charged, their chargers stop energy consumption. Figure 2.9a shows the power profiles described before. In this scenario, the energy consumed to charge the three PEVs (3 kW during 3.2 h for the first PEV, 4.27 h for the second, and 2.67 h for the third) coincides with the residential peak hours. Figure 2.9b shows the corresponding state of charge profiles. As it can be seen, batteries are fully charged in around a third of the time available for the charging period.

For each node, the resulting voltage profiles can be seen on Fig. 2.9c. Comparing with the voltage profiles without PEVs, shown on Fig. 2.7, it can be seen that voltages at nodes 3–8 drop below the established limit of 0.9 p.u., between 18 h and 21 h30 because of the peak of consumption. If this conditions are applied on the grid case 2, PEVs consumption will force voltage profiles to drop even more that those of Figs. 2.8 and 2.9c.

In this case, a single fictive tariff is considered: 1 €/kWh. The resulting cost of charging the three PEVs is 30.4 €. This is proportional to the total amount of energy required to charge the batteries (30.4 kWh, corresponding to 60, 80 and 50% of 16 kWh).

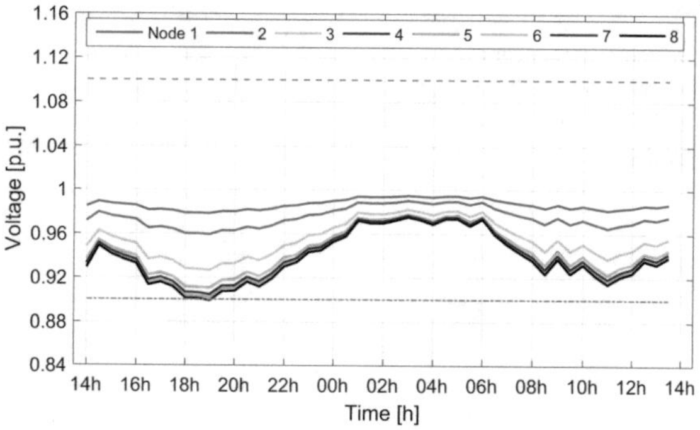

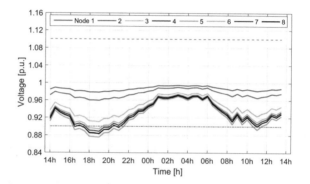

Fig. 2.7 Voltage profiles for the test grid case 1 without connected PEVs. Profiles do not evidence voltage issues. The lower voltage limit (0.9 p.u.) is represented by a dashed line

Fig. 2.8 Voltage profiles for the test grid case 2 without connected PEVs. Certain profiles exceed established limits. The lower voltage limit (0.9 p.u.) is represented by a dashed line

2.4.2 Charging Management—Grid Case 1—Single Tariff

Applying the proposed PEV load management approach to the first grid scenario, the power profiles of Fig. 2.10a are obtained. It can be observed that PEV consumption is redistributed during the whole charging period. It is important to notice that the peak of PEV consumption occurs during the low demand hours. Also, it is important to notice that during base load peak hours, the PEV with the highest initial state of charge injects energy to the grid to compensate the consumption of the other PEVs and maintain the voltages within the limits. The resulting state of charge profiles are shown on Fig. 2.10b. It can be seen that at the end of the charging period, batteries are fully charged. Voltages on all the nodes are kept within the desired limits, even during peak demand hours, as it is shown on Fig. 2.10c.

In this case, the same single fictive tariff is considered: 1 €/kWh. Since the total amount of energy required to fully charge the PEVs is still 30.4 kWh, The optimal cost of charging the PEVs is also 30.4€ as it was for the sub-optimal case.

Fig. 2.9 Case 1 without charging management. **a** Power profiles. **b** State of charge profiles. **c** Voltage profiles

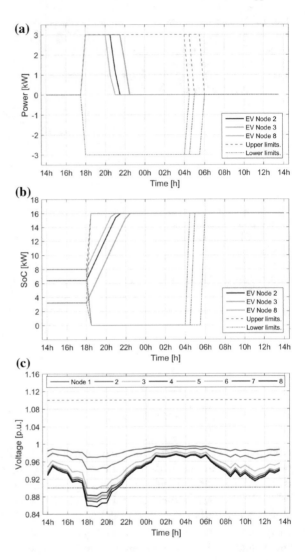

2.4.3 Charging Management—Grid Case 1—Two Tariff Scenario

Now, let us consider a two tariffs scenario: $1 €/$kWh between 22 and 06h, and $1.5 €/$kWh elsewhere. In the sub-optimal case, the consumption is concentrated during the interval of high price, so the cost of the total PEV load without management is $45.6 €$.

Applying the proposed PEV load management approach to the two tariff scenario, results in the power profiles of Fig. 2.11a. Consumption is also redistributed during

Fig. 2.10 Case 1 with charging management and singe tariff. **a** Power profiles. **b** State of charge profiles. **c** Voltage profiles

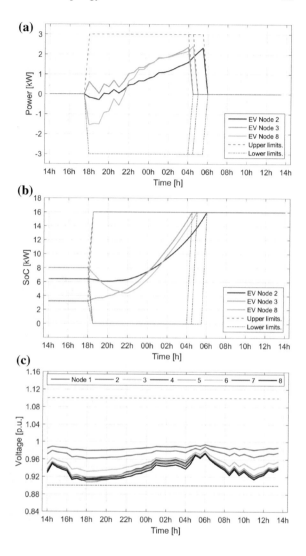

the whole charging period. However, PEVs now try to sell as much of their initial stored energy as possible, during the high-price hours, to reduce the cost of recharging their batteries. State of charge profiles on Fig. 2.11b are useful to verify this behavior. At the end of the high-price hours the three PEVs are fully discharged. They sell their initially stored energy and during the low-tariff hours, they fully charge their batteries. As it is required, the voltage profiles corresponding to each node are kept within the desired limits, even during peak demand hours, as it is shown on Fig. 2.11c.

The optimal cost of charging the PEVs is now reduced to 21.6 €. This cost corresponds to the cost of fully charging the batteries during the low-price hours

Fig. 2.11 Case 1 with
charging management and
two tariffs. **a** Power profiles.
b State of charge profiles. **c**
Voltage profiles

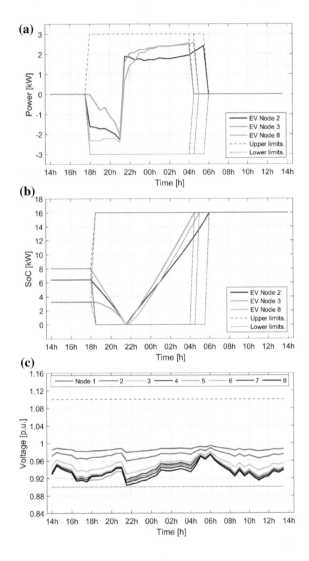

$(3 \times 16\,\text{kWh}\times 1\,\text{€/kWh}= 48.0\,\text{€})$ minus the profit obtained from selling the initially
stored energy during high-tariff hours $(16\,\text{kWh} \times (0.4 + 0.2 + 0.5) \times 1.5\,\text{€/kWh}=26.4\,\text{€})$.

2.4.4 Charging Management—Grid Case 2

Assuming again a single tariff, the proposed approach is tested under the conditions
of the grid case 2. In this case, the voltage profiles do not respect the constraints

Fig. 2.12 Case 2 with
charging management and
single tariff. **a** Power
profiles. **b** State of charge
profiles. **c** Voltage profiles

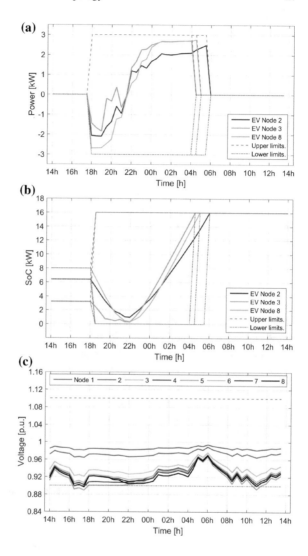

during peak hours, even without PEVs. Given that the peak hours coincide with the
beginning of the charging period, if PEVs do not have any initially stored energy in
their batteries, the constraints of voltage support cannot be satisfied and there is no
feasible solution.

If the initial amount of stored energy is enough, PEVs can provide voltage support
services to the grid by injecting this energy and compensating the base load. Figure 2.8
shows that without PEVs, nodes 4–8 go below the established limits. Applying the
proposed centralized scheduling strategy results in the power profiles of Fig. 2.12a,
and the state of charge profiles of Fig. 2.12b. Here the three PEVs inject energy to
the grid during peak hours, almost until they are fully discharged. After 22 h, when

Table 2.2 IEEE European LV test feeder cable parameters [Iee, Ope]

Type	Number of phases	Positive sequence		Zero sequence	
		R1 [Ω/km]	X1 [Ω/km]	R0 [Ω/km]	X0 [Ω/km]
1	3	3.97	0.099	3.97	0.099
2	3	1.257	0.085	1.257	0.085
3	3	1.15	0.088	1.2	0.088
4	3	0.868	0.092	0.76	0.092
5	3	0.469	0.075	1.581	0.091
6	3	0.274	0.073	0.959	0.079
7	3	0.089	0.0675	0.319	0.076
8	3	0.166	0.068	0.58	0.078
9	3	0.446	0.071	1.505	0.083
10	3	0.322	0.074	0.804	0.093

the base demand decreases, PEVs begin to consume until they are fully charged. It is important to notice that the initial voltage problem of the grid is solved by the support service provided on every node of the grid, as it can be seen on Fig. 2.12c.

In this case, the cost is still 30.4€ as in the case 1, because the total amount of energy required to fully charge the batteries is still 30.4 kWh. It is important to highlight that these optimal schedules are chosen when the solver reaches the stopping criterion. However, given the wide range of freedom for voltage variables when the base demand begins to decrease (especially between 01 h and 05 h), multiple optimal schedules exist, obviously having the same cost.

2.5 Test Case with the IEEE European Low Voltage Test Feeder

In this section, an additional scenario is considered to test the proposed PEV load scheduling method under realistic conditions. In this scenario, a low voltage distribution grid is considered, which is a typical topology for European distribution systems. This radial test grid was published in 2015, by the Test Feeders Working Group of the Distribution System Analysis Subcommittee of the Power Systems Analysis, Computing, and Economics (PSACE) Committee (IEEE). The low voltage test feeder was published to provide a benchmark for researchers studying common European distribution system topologies [Iee, Ope].

2.5.1   The IEEE European Low Voltage Test Feeder and Its Simplification

It is a 240 V nominal line-neutral voltage grid (line-line voltage of 416 V) with a Medium Voltage (11 kV) to Low Voltage transformer at bus 1. The topology, as well as the pre-defined names of some relevant buses, are shown on Fig. 2.13. Most of the terminal buses have single-line loads (a total of 55 loads, marked in this figure). In terms of parameters, all the lines are three-phase lines and have one of the cable characteristics on Table 2.2. For more details on the cable configuration of lines, please check references [Iee, Ope].

The test feeder provides detailed topological coordinates to plot a geographical distribution of the circuit, as it can be observed on Fig. 2.13.

For the purpose of this test, the enumeration of buses is reduced from the original 906 buses, to only 110. These 110 buses are only those where two or more branches diverge, where a load is connected, and where the transformer is connected. The resulting topology and bus enumeration is given on Fig. 2.14, in contrast with the original geographical distribution on the background. Even if the final topology has

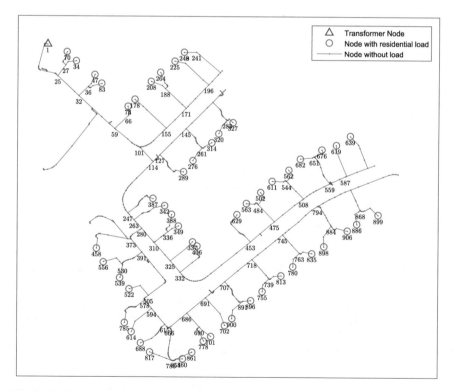

Fig. 2.13   IEEE European low voltage test feeder topology [Iee, Ope]

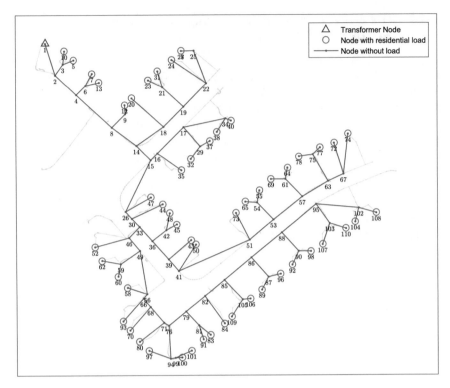

Fig. 2.14 IEEE European low voltage test feeder topology. Simplification of the topology, and enumeration of important nodes for the linear model approximation of this chapter

been reduced, the resistive and reactive values of lines between the mentioned buses are still preserved.[1]

For the linear approximation of this chapter, the reactive positive sequence portions of lines are neglected, as it was previously explained. In this test feeder, the average ratio R/X is 10.2992, while the largest and smallest ratio values are 14.7882 and 1.3185, respectively.

The IEEE European LV test feeder provides 24 h load shapes (in 1 min steps) for each of the 55 loads. These load shapes are series of multipliers applied to a base kW value of load [Ope]. If 1 kW is chosen as the base, the 55 load shapes as well as the sum of these load shapes are shown on Fig. 2.15. These and are available on [Iee] as well. In this figure, the series of multipliers were rearranged to start at 14 h and end at 14 h the next day.[2] In the total load shape it is possible to observe two peaks of consumption, the first one in the morning (around 08h), and a second one, higher in average than the first one, at the beginning of the night (around 19 h).

---

[1] For instance, buses 57 or 101 in this chapter, refer respectively to buses 508 and 861, in the original bus enumeration.

[2] The original series of multiplier have 00 h as start hour and 00 h the next day as end hour [Iee].

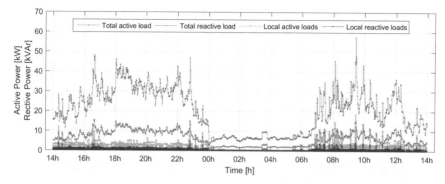

Fig. 2.15 Time series of local and total active and reactive power, available with the IEEE European Low Voltage Test Feeder [Iee, Ope]

Valley hours correspond to the interval between 00 h and 06 h. For the purpose of this chapter, all loads are changed from single to three-phase assigning the original single phase load to each of the phases (total load is multiplied by 3). Furthermore, for the original test feeder, 1 kW is the considered base active power for each load, with a 0.95 power factor, which corresponds approximately to 1.052 kVA. For the purpose of the approximation of this chapter, the 55 loads are considered to have unitary power factor, so the series of multipliers are applied to base loads of 1.052 kW. Furthermore, zero sequence components of cables on Table 2.2 are neglected as well because the system is assumed to be totally balanced.

2.5.2 Comparison Between the Approximated Linear Model and the Accurate Model from OpenDSS

Aiming to validate the linear approximation of this chapter, the loads in the LV test feeder model in *OpenDSS* were modified to be three-phase loads. Moreover, the time step was changed from minutes to half hours, and the corresponding values of the series of multipliers were obtained by averaging over sets of 30 min from the original multipliers. To obtain the reference results from *OpenDSS*, the original base of 1 kW/phase and 0.95 power factor were preserved for all the 55 loads. Total and local load profiles averaged each 30 min (employed on the approximated linear model), are shown on Fig. 2.16.

On Fig. 2.17, a comparison of the magnitude of the voltages at each bus with respect to the distance to the transformer is presented for both the magnitudes obtained with the accurate model of *OpenDSS*, and the linear approximated model of this chapter. This comparison is observed during the peak of average consumption at 18 h, and during the lowest average demand at 00 h, and considering that the transformer

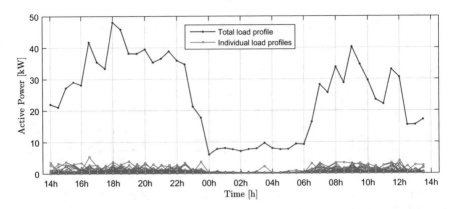

Fig. 2.16 Local and total active power profiles, with time steps of 30 min, for the proposed test scenario

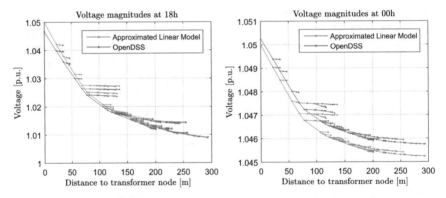

Fig. 2.17 Comparison between voltage profiles obtained with the accurate model of the OpenDSS simulation tool, and the proposed approximated linear model. Comparison for the peak of consumption at 18 h, and the lowest demand at 00 h

voltage is set at 1.05 of the nominal voltage value.[3] During peak hours, it is possible to observe that, as a result of the approximations, the linear model produces errors that decrease as the measurement point goes farther from the transformer. On the other hand, during low demand hours, the error marginally increases with the distance, but it should be noticed that the decay on voltage magnitude is less significant than it is during peak hours. A global point view is captured on the mean squared error (MSE) between voltage magnitudes on each bus during the day ($K = 48$ steps of half hour) given by,

---

[3]To reduce the resulting error comparing with the accurate results of $OpenDSS$, all loads were increased by 5% in the approximated model.

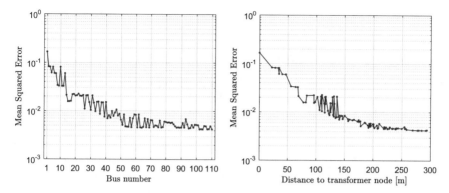

Fig. 2.18 Mean squared error (MSE) of the voltage approximation with respect to buses, and their distances to the transformer. MSE for each bus is computed for a day horizon

$$\text{MSE}_n = \sum_{k=1}^{K} \frac{\left(v_{k_{\text{OpenDSS}}}^n - v_{k_{\text{Approx.}}}^n \right)^2}{K}. \tag{2.24}$$

These measurements are presented on Fig. 2.18. In this figure, the computed errors are organized by buses and in terms of their distance to the transformer bus. Even if it is not the case for all the buses, the tendency of the error is to decrease with the distance to the transformer. For those buses where this does not happen, it should be considered that the error also depends on the load, and each terminal bus has a different series of multipliers and a different behavior in time. It should be mentioned that angles are not considered because reactive portions of loads and lines are neglected in the approximated model (no angular difference between buses in the approximated linear model).

An additional useful illustration is shown on Fig. 2.19. Here, voltage profiles over time are presented for several buses of the grid, on the approximated linear model, and their accurate pairs provided by the *OpenDSS* model. Locating these buses on both Figs. 2.13 and 2.14, it is possible to verify that the error in the voltage magnitude decreases as both the distance to the transformer, and the total load, increase. Finally, voltage magnitude profiles for all the 110 buses, given by the approximated linear model, can be observed on Fig. 2.20.

2.5.3   Description of the Integration of PEVs

After verifying the validity of this model approximation, let us consider a realistic scenario with several PEVs arriving and departing to the grid. On the first place, let us consider several PEV chargers placed on different buses of the grid, with or without residential load. On Fig. 2.21, it is possible to observe where these chargers

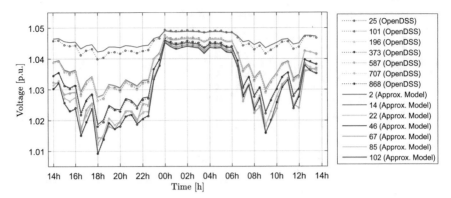

Fig. 2.19 Comparison of voltage day profiles for 7 buses of the grid, obtained with the OpenDSS simulation tool, and the approximated linear model

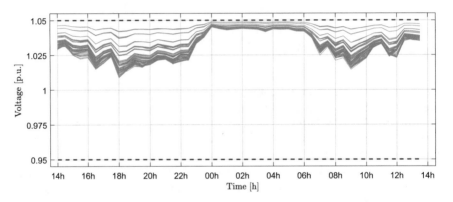

Fig. 2.20 Set of voltage day profiles for all the 110 buses of the reduced grid. Profiles are obtained with the proposed approximated linear model

are randomly placed. In total, there are 33 chargers with nominal powers of 3.3 and 7.5 kW.

In this test, PEVs arrive in a random way, following a *Poisson* model with variable rate of arrivals according to the hour, and variable connection times. The highest rate of arrivals is 10 PEVs/h at 17 h and it decays up to 0.5 PEVs/h at 16 h the next day. The number of arrivals and departures each half hour, can be checked on Fig. 2.22, in contrast with the total load profile. It can be observed that most of the PEVs arrive during peak of consumption in the afternoon. Vehicle battery capacities can be 8.8 kWh with a probability of 30%, and 20 kWh with a probability of 70%. Furthermore, chargers can have a limit of power of 3.3 kW with a probability of 80%, and 7.4 kW with a probability of 20%. To reduce the impact on the battery lifespan, states of charge are limited to 25–85% for 8.8 kWh capacities (2.2–7.5 kWh), and 30–80% for the 20 kWh capacities (6–16 kWh). A summary of these assumptions and the evaluated scenarios of this section can be found on Table 2.3. On Fig. 2.23,

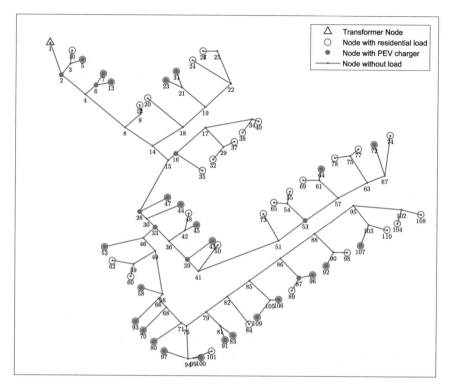

Fig. 2.21 Random allocation of PEV chargers on the buses of the simplified version of the IEEE European low voltage test feeder

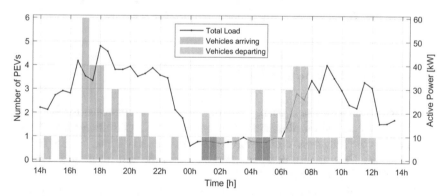

Fig. 2.22 Profiles of arrivals and departures of PEVs during the test day, in contrast with the total load profile of the grid

it is possible to observe a contrast between the arrivals and departures of PEVs and the amount con connected PEVs in total, per nominal power of the charger, and per battery capacity.

Table 2.3 Descriptive summary and assumptions of the considered study cases

Item	Description
Chargers	3.3 kW with probability of 80%
	7.5 kW with probability of 20%
Batteries	8.8 kWh with probability of 30%
	20 kWh with probability of 70%
Constraints on batteries	Between 25 and 85% for 8.8 kWh
	Between 30 and 80% for 20 kWh
Highest rate of arrivals	10 PEVs/h at 17 h
Lowest rate of arrivals	0.5 PEVs/h at 16 h next day
Peak of connected PEVs	30 PEVs at 01 h in the morning
Distribution system info.	IEEE European low voltage test feeder
Evaluated scenarios	Single and Double tariff
	with variations on voltage constraints
Time period	24 h (half hour steps)

It is possible to observe that there is a large amount of PEVs connected during low demand hours. Moreover, as it can be expected from the previously mentioned probabilities, the most common nominal power of chargers of connected PEVs is 3.3 kW, while most of the PEVs have batteries of 20 kWh.

Based on these profiles of PEVs arriving and departing, and the approximated linear modeling of the low voltage grid, multiple alternatives were considered for the cost function to minimize, and the voltage constraints, in the proposed linear PEV load scheduler. Several cases are explained in following sections.

2.5.4 Results with Single Tariff

In this first scenario, a single energy consumption tariff is employed for the whole day. In such a case, all the alternative consumption profiles fulfilling the whole set of constraints have the same cost, because consuming energy at one or another hour costs the same. This implies that the minimization of the cost function is achieved by any set of schedules satisfying the constraints.

Considering bidirectional chargers, and voltage limits for all the buses between $v_{min} = 0.95V_{nom}$ and $v_{max} = 1.05V_{nom}$, at all the time steps, the scheduling results provided by the proposed approach, are presented on Fig. 2.24.

As it can be observed in these figures, even if PEV load is added during the original peak of consumption at 18 h, voltages at all the buses remain inside the range of the constraints because the voltage at the transformer is set at $1.05V_{nom}$. Moreover, it can be observed that a big portion of the PEV load is consumed between 04 h and 12 h, because PEV owners choose their departure times in this range of hours. This

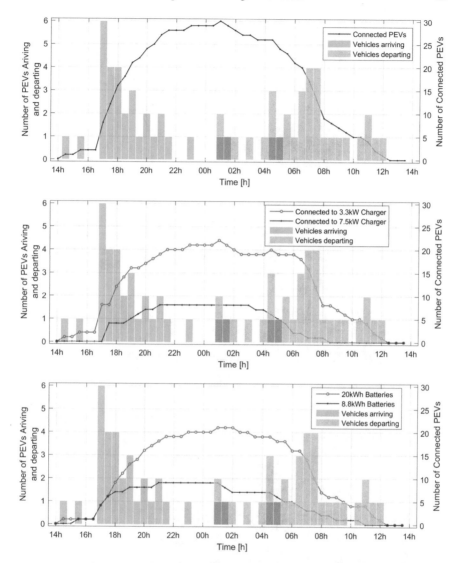

Fig. 2.23 Profiles of arrivals and departures of PEVs during the test day, in contrast with the resulting total amount of connected PEVs, the total amount of connected PEVs to 3.3 and 7.5 kW chargers, and the total amount of PEVs with battery capacities of 20 and 8.8 kWh

behavior is beneficial for the grid because the PEV consumption is naturally allocated during valley hours. However, it is not the general case since owners may choose departure times coincident with peak hours, or just after peak hours, as well.

To force the scheduler to avoid PEV load allocation during peak hours, without including second tariff incentives, it is possible to tight voltage constraints. Such case is presented on Fig. 2.25, where voltages at all the buses during the day, are

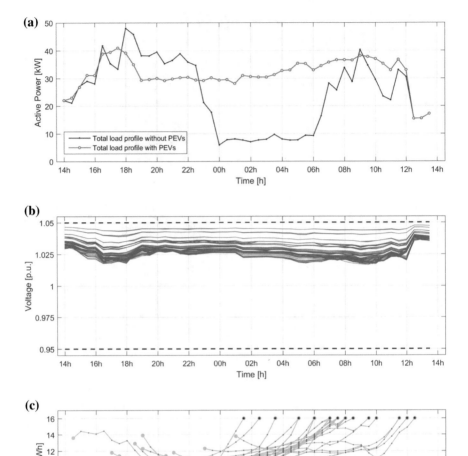

Fig. 2.24 Scenario with single tariff, and regular voltage constrains. **a** Total load profiles with and without PEV load; **b** Resulting voltage profiles with PEVs after optimal scheduling; **c** Resulting optimally scheduled state of charger profiles

constrained to be between $v_{min} = 1.01V_{nom}$ and $v_{max} = 1.05V_{nom}$. As a consequence, the load does not exceed 50 kW at any hour during the day, and more PEV load is forced to be consumed during valley hours. Voltages at 18 h, at 07 h, and at 11 h reach values closer to the lower boundary, forcing PEV load to be allocated elsewhere in

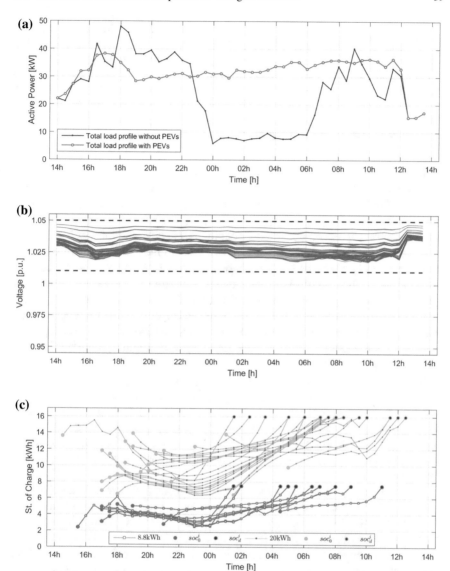

Fig. 2.25 Scenario with single tariff, and more restrictive voltage constrains. **a** Total load profiles with and without PEV load; **b** Resulting voltage profiles with PEVs after optimal scheduling; **c** Resulting optimally scheduled state of charger profiles

time. Moreover, in both cases it should be noticed that PEVs reach their desired states of charge at the programmed disconnection times, and none of the state of charge profiles exceeds the allowed limits.

Voltages limits can be defined even more restrictively. Nevertheless, the problem may become unfeasible depending on the departure times of PEVs, and their available stored energy. Moreover, using single tariffs, the potential of bidirectional chargers is not properly exploited as it can be observed.

2.5.5 Results with Double Tariff

Now, let us consider two different tariffs, one for day hours and a smaller one for night hours. The low tariff is applied between 22 h and 06 h in the morning the next day. First let us consider unidirectional chargers, and constraints of voltages between $v_{min} = 0.95 V_{nom}$ and $v_{max} = 1.05 V_{nom}$. The scheduling results provided by the proposed approach under these conditions, are presented on Fig. 2.26.

In this case, it is important to observe that most of the PEV load is forced to be consumed during low-tariff hours because the total cost of energy can be reduced. However, given this incentive to consumption, it is likely for PEVs to allocate great quantities of load in this interval, without any uniformity, as long as voltage constraints on all buses are fulfilled. As a consequence, strong peaks in consumption may occur as it can be observed at the beginning and the end of the low-tariff period. It is here where more restrictive voltage limits are very useful to force PEV load to be allocated more evenly during the whole period of low tariff.

More restrictive limits can be proposed during low-tariff hours to avoid this strong peaks of PEV consumption. Now, let us constrain voltages between $v_{min} = 1.02 V_{nom}$ and $v_{max} = 1.05 V_{nom}$, only during low-tariff hours. The scheduling results provided by the proposed approach under these conditions, are presented on Fig. 2.27. It can be observed that the strong peaks at the beginning and the end of the low-tariff period are cleared and PEV load is forced to be allocated elsewhere, during the low-tariff interval as well.

In both Figs. 2.26 and 2.27, it can be observed again that state of charge profiles reach the desired state of charge at the programmed disconnection times, and none of them exceeds the allowed limits. Moreover, as it was established, batteries are never discharged throughout the time, given the unidirectionality of chargers.

Now, let us consider bidirectional chargers, and the original voltage constraints again, between $v_{min} = 0.95 V_{nom}$, and $v_{max} = 1.05 V_{nom}$. In this case, batteries consider discharging during high-tariff hours to get a profit (assuming it can be payed by the utility grid company). Batteries are discharged as much as possible before the low-tariff hours, and then PEVs consume as much as possible, such that the total cost of charging is the lowest possible. As a result PEVs have lower initial state of charge at the beginning of low-tariff hours, which means they have a larger energy need. This greediness effect can be counterproductive since stronger peaks, than those with unidirectional chargers, are likely to happen at the beginning and end of the low-tariff hours, as it can be observed on Fig. 2.28. Furthermore, since PEVs inject energy to the grid during high-tariff hours, the potential of bidirectional chargers is not exploited properly. Load is simply reallocated without cutting peaks or filling the

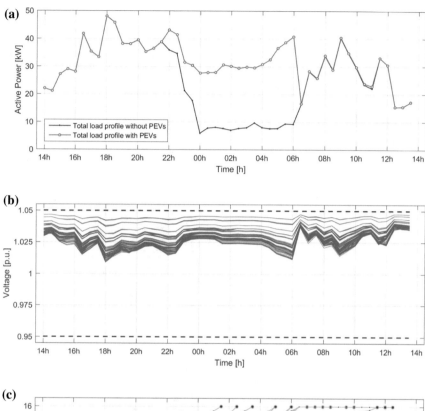

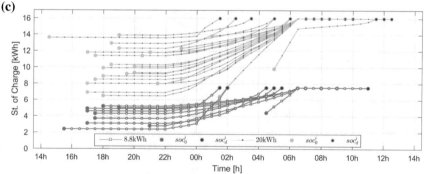

Fig. 2.26 Scenario with double tariff, unidirectional chargers, and regular voltage constrains. **a** Total load profiles with and without PEV load; **b** Resulting voltage profiles with PEVs after optimal scheduling; **c** Resulting optimally scheduled state of charger profiles

valley hours. Again, it is here where more restrictive voltage limits are very useful to force PEV load to be allocated more evenly during the day.

Figure 2.29 shows the scheduling results provided by the proposed approach when voltage is constrained between $v_{min} = 1.03V_{nom}$, and $v_{max} = 1.05V_{nom}$, during

(a)

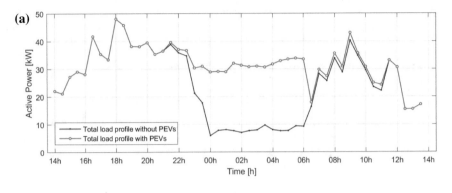

(b)

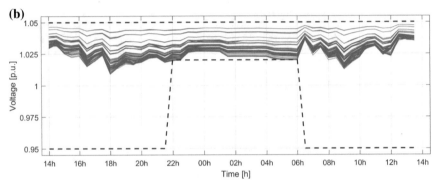

(c)

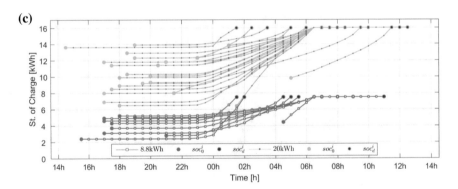

Fig. 2.27 Scenario with double tariff, unidirectional chargers, and more restrictive voltage constrains during low-tariff hours. **a** Total load profiles with and without PEV load; **b** Resulting voltage profiles with PEVs after optimal scheduling; **c** Resulting optimally scheduled state of charger profiles

low-tariff hours, and between $v_{min} = 1.01V_{nom}$, and $v_{max} = 1.05V_{nom}$ elsewhere. These constraints put boundaries on the greedy effect of the double tariff, and force PEVs to reduce the amount of energy they allocate during low-tariff hours. As it can be observed, PEVs take as much advantage of the low-tariff as they can since most

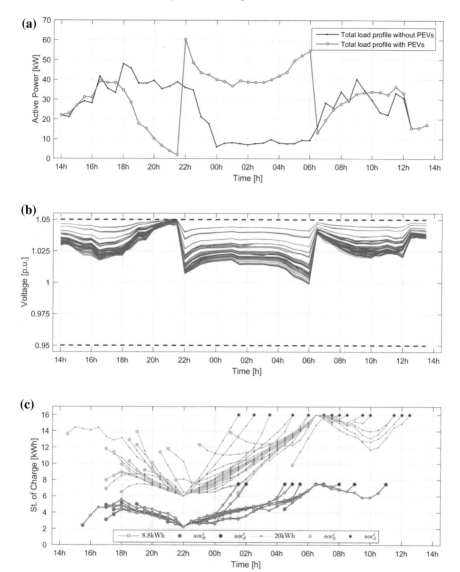

Fig. 2.28 Scenario with double tariff, bidirectional chargers, and regular voltage constrains. **a** Total load profiles with and without PEV load; **b** Resulting voltage profiles with PEVs after optimal scheduling; **c** Resulting optimally scheduled state of charger profiles

of the voltages are on the edge of the lower constraint during this period. On the other hand, constraints during high-tariff hours have the effect of forcing PEVs to avoid load allocation during peaks of consumption, and even reallocate those peaks using the batteries. Consequently, it can be observed that the original peak at 18 h, is reduced and it becomes similar to the peaks of consumption after 06 h.

(a)

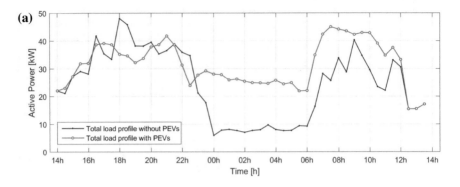

(b)

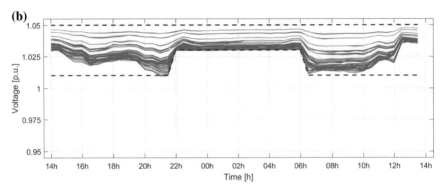

(c)

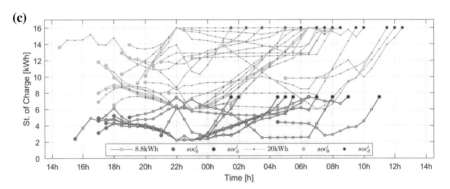

Fig. 2.29 Scenario with double tariff, bidirectional chargers, and more restrictive voltage constrains during both low and high-tariff hours. **a** Total load profiles with and without PEV load; **b** Resulting voltage profiles with PEVs after optimal scheduling; **c** Resulting optimally scheduled state of charger profiles

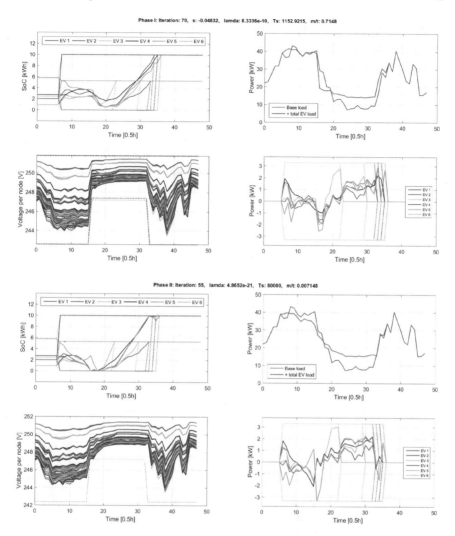

Fig. 2.30 Final iteration for Phase I and Phase II routines in the last example of Sect. 2.5.5 for vehicles that arrive at 17 h

2.6   Matlab Scripts

The following Matlab scripts allow you to check the first example of this chapter (Sect. 2.4). To be used as a templates for other languages, these routines are implemented using mostly basic Matlab functions. The reader is encourage to modify this functions and explore different scenarios.

The script on Sect. 2.6.1 defines the conditions of the example as it was described in Sect. 2.4. Then, the function of Sect. 2.6.2 is called to compute an initial guess for the PEV charging/discharging schedules such that the whole set of constraints is satisfied. Finally, a Phase II routine is provided on Sect. 2.6.3. This routine optimizes the objective function (cost of charging the PEVs) starting from the initial guesses provided by the Phase I routine.

Phase I and II routines are implemented based on the descriptions of *Newton's method with equility constraints*, and the *Barrier method* from [BV04].

It is important to mention that in these scripts, a variable called watch_evolution can be set to 1 such that the optimization routines plot the evolution of the variables of interest. For instance, the final iterations of the Phase I and Phase II routines for the last example on Sect. 2.5.5 are shown on Fig. 2.30. These illustrations correspond to the outcomes of the routines for vehicles arriving at 17 h in that example.

2.6.1 Study Case Definition Script

As it was mentioned above, the following script defines the conditions of the example in Sect. 2.4. Once these conditions are defined, the optimal centralized scheduling routines (Phases I and II) are called.

```
clear
clc
close all

% Load profiles for the 8 nodes of the test

load_nodes = ...
[1.99 1.4 1.74 1.77 1.96 2.51 2.5 2.44 2.87 2.8 2.82 2.63 2.69 2.36 2.45...
2.39 1.87 1.88 1.64 1.58 1.31 1.26 0.77 0.94 0.81 0.88 0.72 0.91 0.87 ...
0.75 0.84 1.01 0.74 1.2 1.51 1.67 1.86 2.23 1.8 2.15 2.03 2.17 2.43 ...
2.27 2.15 1.94 1.96 1.69;
2.05 1.52 1.8 1.83 1.88 2.48 2.34 2.57 2.84 2.83 2.96 2.59 2.7 2.45 ...
2.47 2.32 1.9 1.9 1.61 1.54 1.31 1.18 0.87 0.75 0.8 0.81 0.65 0.76 0.87 ...
0.78 0.64 0.96 0.85 1.24 1.41 1.73 1.79 2.2 1.69 2.22 1.95 2.14 2.49 ...
2.22 2.13 1.84 1.93 1.68;
2.01 1.5 1.6 1.8 1.87 2.41 2.41 2.5 2.84 2.78 2.87 2.57 2.62 2.37 2.48 ...
2.26 2.04 1.89 1.68 1.48 1.39 1.23 0.9 0.82 0.87 0.88 0.63 0.74 0.84 ...
0.72 0.77 1.01 0.77 1.11 1.45 1.69 1.82 2.16 1.73 2.21 1.9 2.12 2.48 ...
2.34 2.15 1.87 1.94 1.69;
2.14 1.42 1.69 1.87 1.9 2.47 2.52 2.47 2.92 2.82 2.77 2.66 2.51 2.39 ...
2.59 2.4 2 1.96 1.61 1.67 1.3 1.17 0.83 0.89 0.81 0.86 0.71 0.82 0.89 ...
0.89 0.75 0.86 0.83 1.23 1.36 1.62 1.79 2.25 1.75 2.32 1.94 2.26 2.45 ...
2.28 2.21 1.86 1.98 1.87;
1.97 1.49 1.71 1.85 1.96 2.54 2.42 2.66 2.78 2.94 2.9 2.66 2.66 2.42 ...
2.48 2.25 1.89 1.88 1.65 1.63 1.37 1.25 0.75 0.88 0.89 0.78 0.79 0.84 ...
0.98 0.74 0.69 0.92 0.72 1.28 1.4 1.63 1.83 2.15 1.73 2.07 1.88 2.28 ...
2.54 2.22 2.12 1.89 1.92 1.71;
2.02 1.46 1.78 1.73 1.96 2.56 2.35 2.59 2.72 2.89 2.94 2.65 2.71 2.48 ...
2.49 2.28 2.09 1.81 1.53 1.57 1.34 1.15 0.83 0.8 0.94 0.8 0.76 0.78 ...
0.91 0.78 0.75 0.96 0.81 1.18 1.47 1.66 1.84 2.11 1.77 2.16 1.88 2.18 ...
2.45 2.18 2.22 1.85 1.88 1.77;
```

```
  2.03 1.46 1.61 1.85 1.93 2.52 2.4 2.45 2.86 2.88 2.9 2.66 2.75 2.4 2.52 ...
  2.23 1.96 1.87 1.68 1.65 1.37 1.31 0.76 0.86 0.89 0.75 0.63 0.76 0.85 ...
  0.73 0.78 0.95 0.78 1.16 1.41 1.6 1.81 2.23 1.86 2.14 1.96 2.3 2.43 ...
  2.36 2.05 1.79 1.98 1.79;
  2.07 1.46 1.65 1.82 1.92 2.41 2.41 2.55 2.82 2.74 2.88 2.69 2.54 2.36 ...
  2.59 2.41 2.05 1.94 1.61 1.52 1.38 1.21 0.79 0.95 0.85 0.76 0.75 0.78 ...
  0.93 0.71 0.78 0.91 0.68 1.07 1.42 1.6 1.87 2.18 1.84 2.19 2.01 2.24 ...
  2.4 2.22 2.21 1.81 1.92 1.73];


% A_tilde matrix for the system in the first case
A_tilde_1 = ...
  [-40 20 0 0 0 0 0 0;
   20 -30 10 0 0 0 0 0;
   0 10 -26.666666666666668 6.666666666666667 0 10 0 0;
   0 0 6.666666666666667 -13.333333333333334 6.666666666666667 0 0 0;
   0 0 0 6.666666666666667 -6.666666666666667 0 0 0;
   0 0 10 0 0 -30 20 0;
   0 0 0 0 0 20 -30 10;
   0 0 0 0 0 0 10 -10];

% A_tilde matrix for the system in the second case
A_tilde_2 = ...
  [-40 20 0 0 0 0 0 0;
   20 -26.666666666666668 6.666666666666667 0 0 0 0 0;
   0 6.666666666666667 -20.666666666666668 4 0 10 0 0;
   0 0 4 -10.666666666666668 6.666666666666667 0 0 0;
   0 0 0 6.666666666666667 -6.666666666666667 0 0 0;
   0 0 10 0 0 -30 20 0;
   0 0 0 0 0 20 -30 10;
   0 0 0 0 0 0 10 -10];

% input voltage and resistance for the transformer
v_0 = 230;
R_0 = 0.05;

% voltage constraints for each node of the grid
v_max = zeros(size(load_nodes))+v_0*1.1;
v_min = zeros(size(load_nodes))+v_0*0.9;

% fictive tariff for 06h-23h (day tariff)
tariffs = ones(1,length(load_nodes))*1.5;
% fictive tariff for 23h01-05h59 (night tariff)
tariffs(16:32) = 1;

% System matrix A and transformer voltage vector v_k_0 from modeling
v_k_0 = [-v_0; 0; 0; 0; 0; 0; 0; 0];
A = A_tilde_1\eye(length(A_tilde_1));
% A = A_tilde_2\eye(length(A_tilde_2)); % uncomment to check Case 2

% voltage imposed on each node by the transformer and the system topology
v_k_base = repmat((1/R_0)*A*v_k_0,1,length(load_nodes))...
    +(1/v_0)*A*(load_nodes*1000);

% Parameters for each of the 3 PEVs
soc_cap = [16 16 16]';    % storage capacities [kWh]
soc_d = [16 16 16]';      % desired final states of charge [kWh]
soc_0 = [6.4 3.2 8]';     % initial states of charge [kWh]
p_max_c = [3 3 3]';       % max charging rates [kW]
```

```
p_max_i = [3 3 3]';             % max discharging rates [kW]
Ki = [24 21 22]';               % final time step of connection periods
dt = 0.5;                       % time step [h]
t_0 = 8;                        % t_0+dt is the time step under evaluation
                                % t_0 is the time step corresponding to soc_0
                                % (vehicles can be connected from before but in
                                % this case this is the initial time step for
                                % the three EVs)
nodes = [2 3 8]';               % Nodes where EVs are connected

Lhat_ = sum(load_nodes);        % Total load profile (neglecting losses)

watch_evolution = 1; % Set to 1 in order to watch the routines' evolution

[pow_traj_ast,s] = Book_EV_Centralized_Scheduling_PhaseI(soc_cap,soc_d,...
                soc_0,p_max_c,p_max_i,t_0,Ki,dt,A,v_0,v_k_base,v_max,...
                v_min,Lhat_',nodes,watch_evolution);
if s > 0 % if the maximum infeasibility goes below zero (all the voltage
         % constraints are feasible and the Phase II problem can find the
         % optimal schedules)
else
    [pow_traj_ast,soc_traj_ast] = Book_EV_Centralized_Scheduling_PhaseII...
        (soc_cap,soc_d,soc_0,p_max_c,p_max_i,t_0,Ki,dt,A,v_0,v_k_base,...
        v_max,v_min,Lhat_',nodes,tariffs,pow_traj_ast,watch_evolution);
end
```

2.6.2 Phase I Optimisation Routine

This routine is called to compute an initial guess for the PEV charging/discharging schedules such that the whole set of constraints is satisfied. In this case, state of charge, and power constraints are considered *inflexible* constraints. On the other hand, voltage constraints are considered to be flexible. Keeping this in mind, this routine acts as a Phase I optimization routine minimizing the bound on the maximum in-feasibility of voltage constraints [BV04]. If it is not possible to find initial guesses for the charging/discharging schedules satisfying the whole set of constraints, then this function returns the best possible schedules satisfying at least the state of charge and power constraints. If the bound on the maximum in-feasibility of voltage constraints goes below zero, it means that the whole set of constraints is feasible, then the schedules defined by the Phase I routine are assigned as the initial guess for the next routine (Phase II).

```
function [pow_traj_ast,s] = ...
    Book_EV_Centralized_Scheduling_PhaseI(soc_c,soc_d,soc_0,p_max_c,...
    p_max_i,t_0,Ki,dt,A,v_0,v_k_base,v_max,v_min,Lhat,nodes,...
    watch_evolution)

% ------ Description inputs
% soc_c : battery available capacities [kWh]
% soc_d : final desired states of charge [kWh]
% soc_0 : initial states of charge [kWh]
% p_max_c : max charging rates [kW]
```

```
% p_max_i : max discharging rates [kW]
% t_0 : time step corresponding to soc_0
% Ki : total time steps
% dt : time step
% A : system matrix
% v_0 : transformer voltage
% v_k_base : voltages imposed by the transformer and base loads on the grid
%            nodes
% v_max : upper boundary for voltages on each node
% v_min : lower boundary for voltages on each node
% Lhat_ : total load forecast on the transformer
% nodes : nodes of the grid where EVs are connected
% watch_evolution : (1) plots internal evolution, (otherwise) ignores

% ------ Description outputs
% pow_traj_ast : optimal power trajectory
% s : max infeasibility bound on the voltage constraints

% ------ Onjective
% Drive the bound "s" bellow zero in order to provide an initial guess for
% the charging schedules such that the voltage, soc, and power constraints
% are respected
% (Minimize s)

% Initialization of variables

% Number of PEVs
N = length(Ki);
Lhat_=Lhat';

% construction of soc_k constraints for visualization of evolution
soc_min = zeros(N,length(Lhat_)+1);
soc_min(sub2ind(size(soc_min),(1:N)',Ki*0+1)) = soc_0-1e-3;
soc_min(sub2ind(size(soc_min),(1:N)',Ki*0+t_0+2)) = -soc_0+2e-3;
soc_min(sub2ind(size(soc_min),(1:N)',Ki+t_0+1)) = +soc_d-2e-3;
soc_min = cumsum(soc_min,2);
soc_min=soc_min(:,2:end);

soc_max = zeros(N,length(Lhat_)+1);
soc_max(sub2ind(size(soc_max),(1:N)',Ki*0+1)) = soc_0+1e-3;
soc_max(sub2ind(size(soc_max),(1:N)',Ki*0+t_0+2)) = -soc_0+soc_c-2e-3;
soc_max(sub2ind(size(soc_max),(1:N)',Ki+t_0+1)) = -soc_c+soc_d+2e-3;
soc_max = cumsum(soc_max,2);
soc_max=soc_max(:,2:end);

% construction of soc_k constraints for the routine
soc_lo = zeros(N,length(Lhat_)+1);
soc_lo(sub2ind(size(soc_lo),(1:N)',Ki*0+1)) = +1e-3;
soc_lo(sub2ind(size(soc_lo),(1:N)',Ki*0+t_0+2)) = -soc_0-2e-3;
soc_lo(sub2ind(size(soc_lo),(1:N)',Ki+t_0+1)) = Inf;
soc_lo = cumsum(soc_lo,2);
soc_lo=soc_lo(:,2:end);
soc_lo(:,1:t_0)=[];
soc_lo(:,max(Ki):end)=[];
soc_lo=soc_lo(:);
soc_lo(soc_lo==Inf)=[];

soc_up = zeros(N,length(Lhat_)+1);
soc_up(sub2ind(size(soc_up),(1:N)',Ki*0+1)) = -1e-3;
```

```
soc_up(sub2ind(size(soc_up),(1:N)',Ki*0+t_0+2)) = -soc_0+soc_c+2e-3;
soc_up(sub2ind(size(soc_up),(1:N)',Ki+t_0+1)) = Inf;
soc_up = cumsum(soc_up,2);
soc_up=soc_up(:,2:end);
soc_up(:,1:t_0)=[];
soc_up(:,max(Ki):end)=[];
soc_up=soc_up(:);
soc_up(soc_up==Inf)=[];

% construction of charging rate constraints for each EV
u_up = zeros(N,length(Lhat_));
u_up(sub2ind(size(u_up),(1:N)',Ki*0+t_0+1*(t_0==0))) = -p_max_c*(t_0>0);
u_up(sub2ind(size(u_up),(1:N)',Ki+t_0)) = p_max_c;
u_up = cumsum(u_up,2,'reverse')+1e-3;

u_lo = zeros(N,length(Lhat_));
u_lo(sub2ind(size(u_lo),(1:N)',Ki*0+t_0+1*(t_0==0))) = +p_max_i*(t_0>0);
u_lo(sub2ind(size(u_lo),(1:N)',Ki+t_0)) = -p_max_i;
u_lo = cumsum(u_lo,2,'reverse')-1e-3;

% construction of initial guess of charging rates
% note: this Phase I initial guess is defined by fastest charing schedules
p_max_until = floor((soc_d-soc_0)./(p_max_c*dt));
p_to_goal = ((soc_d-soc_0)./(p_max_c*dt)-p_max_until).*p_max_c;
u = zeros(N,length(Lhat_)+1);
u(sub2ind(size(u),(1:N)',p_max_until+t_0+1)) = p_max_c;
u(sub2ind(size(u),(1:N)',p_max_until*0+t_0+1)) = ...
    u(sub2ind(size(u),(1:N)',p_max_until*0+t_0+1))- p_max_c;
u = cumsum(u,2,'reverse');
u(sub2ind(size(u),(1:N)',p_max_until+t_0+2)) = p_to_goal;
u=u(:,2:end);

% construction of a mask to avoid influence on unexisting controls
mask = zeros(N,length(Lhat_)+1);
mask(sub2ind(size(mask),(1:N)',Ki+t_0+1)) = 1;
mask(sub2ind(size(mask),(1:N)',Ki*0+t_0+1)) = -1;
mask = cumsum(mask,2,'reverse');
mask=mask(:,2:end);
mm =(mask(:)==1)';

% constraints on A_ev*u defined by constraints on voltages of each node
v_up = (v_0/1000)*(v_max(:,sum(mask,1)>0)-v_k_base(:,sum(mask,1)>0))+1e-3;
v_lo = (v_0/1000)*(v_min(:,sum(mask,1)>0)-v_k_base(:,sum(mask,1)>0))-1e-3;

% construction of equality constraints matrix "A_soc_end" for the final
% states of charge
A_soc_end = zeros(N,N*max(Ki));
for tt = 1 : max(Ki)
    A_soc_end(:,(tt-1)*N+1:(tt)*N) = diag(mask(:,t_0+tt))*dt;
end

% construction of inequality constraints matrix "A_soc" for the partial
% states of charge at each time step "k"
A_soc = repmat(A_soc_end,max(Ki)-1,1);
for nn = 1 : max(Ki)-1
    A_soc((nn-1)*N+1:nn*N, (nn)*N+1:end) = ...
        0*A_soc((nn-1)*N+1:nn*N, (nn)*N+1:end);
end
```

```
Indices=zeros(1,sum(max(Ki)-Ki));
prev_count = 0;
for ii = 1:length(Ki)
    Indices_Ki = (Ki(ii)):(max(Ki)-1);
    Indices(prev_count+1:prev_count+length(Indices_Ki)) = ...
        (Indices_Ki-1)*N+ii;
    prev_count = prev_count +length(Indices_Ki);
end
A_soc(Indices,:)=[];
A_soc(:,sum(A_soc_end)==0)=[];
A_soc_end(:,sum(A_soc_end)==0)=[];
A_soc_end(:,end+1) = A_soc_end(:,end)*0;


% Elimination of nodes (columns of the system matrix A) without PEVs.
A_ev = A(:,nodes);
% Repetition of "A_ev" matrix for each time step until the disconnection of
% the last PEV.
A_ev_re = zeros(length(A)*max(Ki),sum(Ki));
prev_row = 0;
prev_col = 0;
for tt = 1:max(Ki)
    N_connected = sum(mask(:,t_0+tt));
    A_ev_re(prev_row+1:prev_row+length(A),...
            prev_col+1:prev_col+N_connected) ...
                        = A_ev(:,mask(:,t_0+tt)==1);
    prev_row = prev_row+length(A);
    prev_col = prev_col+N_connected;
end

% Partial state of charge inequality constraint matrices for lower and
% boundaries, including the max infeasibility bound "s" variable
% - For states of charge constraints (no influence of "s"):
A_soc_s_lo = [A_soc zeros(sum(Ki-1),1)];
A_soc_s_up = [A_soc -zeros(sum(Ki-1),1)];

% - For charging rates constraints (no influence of "s"):
A_u_s_lo = [eye(sum(Ki)) zeros(sum(Ki),1)];
A_u_s_up = [eye(sum(Ki)) -zeros(sum(Ki),1)];

% - For voltage constraints:
A_v_s_lo = [A_ev_re ones(size(A_ev_re,1),1)];
A_v_s_up = [A_ev_re -ones(size(A_ev_re,1),1)];

% Parameters of descent, and backtracking line search
alpha = 0.5;
beta = 0.8;
mu = 10;
t = 1;
t_pre = t;
num_constraints = 2*sum(Ki)+2*size(A_soc,1)+2*size(A_ev_re,1);
Epsilon_lambda = 1e-20;
Epsilon_Barrier = 1e-50;

% Max number of iterations
Max_Iter=1000;

for ii = 1:Max_Iter
```

```
% computes the states of charge trajectories
x_k = A_soc*u(mm)';

% computes voltage contribution of EVs
v = A_ev_re*u(mm)';

if ii == 1 % initialize s
    s = 1e-3 + max([-v+v_lo(:); v-v_up(:)]);
end

% computes the extended objective function's current value (including
% inequalities implicitly)
H = t*s...
    - sum(log(-v_lo(:)+v+s)+log(v_up(:)-v+s))...
    - sum(log(-soc_lo+x_k)+log(soc_up-x_k))...
    - sum(log(-u_lo(mm)'+u(mm)')+log(u_up(mm)'-u(mm)'));

% computes the gradient of the extended objective function
dH_dus = A_v_s_lo'*(1./(v_lo(:)-v-s))...
        + A_v_s_up'*(1./(v_up(:)-v+s))...
        + A_soc_s_lo'*(1./(soc_lo-x_k))...
        + A_soc_s_up'*(1./(soc_up-x_k))...
        + A_u_s_lo'*(1./(u_lo(mm)'-u(mm)'))...
        + A_u_s_up'*(1./(u_up(mm)'-u(mm)'));

dH_dus(end) = dH_dus(end)+t;

% computes de Hessian given the implicit inclusion of the inequalities
% in the objective function
I_x_lo = diag(1./(soc_lo-x_k).^2);
I_x_up = diag(1./(soc_up-x_k).^2);

I_u_lo = diag(1./(u_lo(mm)'-u(mm)').^2);
I_u_up = diag(1./(u_up(mm)'-u(mm)').^2);

I_v_lo = diag(1./(v_lo(:)-v-s).^2);
I_v_up = diag(1./(v_up(:)-v+s).^2);

Hessian =  + A_v_s_lo'*I_v_lo*A_v_s_lo + A_v_s_up'*I_v_up*A_v_s_up ...
           + A_soc_s_lo'*I_x_lo*A_soc_s_lo ...
           + A_soc_s_up'*I_x_up*A_soc_s_up ...
           + A_u_s_lo'*I_u_lo*A_u_s_lo + A_u_s_up'*I_u_up*A_u_s_up;

% normalizaion of the gradient
dH_dus_normalized = dH_dus/norm(dH_dus);

% computes the direction of descent in the Newton's method with
% equality constraints
if rcond([Hessian A_soc_end';A_soc_end zeros(size(A_soc_end,1))])>1e-15
    Delta_x_w = ...
        -[Hessian A_soc_end';A_soc_end zeros(size(A_soc_end,1))]\...
        [dH_dus_normalized;zeros(size(A_soc_end,1),1)];
else % If the extended matrix in the equality constrained
     % problem is badly conditionned, the routine stops
    break
end

% Extracts the elements that correspond to the direction of descent
Delta_x = Delta_x_w(1:length(dH_dus_normalized));
```

```matlab
% w = Delta_x_w(length(dH_dus_normalized)+1:end);

% Computes the decrement
lambda = (Delta_x'*Hessian*Delta_x)^2;

% If the decrement is less that the threshold after changing "t",
% the routine stops
if t > t_pre && lambda < Epsilon_lambda
    break
else
end
t_pre = t;

% Initial step size in the backtracking line search
Ts = 100000;

H_ir = Inf;

% backtracking line search subroutine
while (H_ir > (H+alpha*Ts*(dH_dus_normalized'*Delta_x)))
    u_ir = u;
    u_ir(mm) = u(mm) + Ts*Delta_x(1:end-1)';
    s_ir = s + Ts*Delta_x(end);
    % computes the possible state of charge trajectories

    x_k_ir = A_soc*u_ir(mm)';

    % computes possible voltage contribution of EVs
    v_ir = A_ev_re*u_ir(mm)';

    % checks is possible variable values lie within specified ranges
    if (sum(v_ir < (v_lo(:)-s_ir))+ sum(v_ir > (v_up(:)+s_ir))+ ...
            sum(x_k_ir < soc_lo) + sum(x_k_ir > soc_up) + ...
            sum(u_ir(mm)' < u_lo(mm)') + sum(u_ir(mm)' > u_up(mm)'))>0
        Ts=beta*Ts;
    else
        H_ir = t*s_ir...
            - sum(log(-v_lo(:)+v_ir+s_ir)+log(v_up(:)-v_ir+s_ir))...
              - sum(log(-soc_lo+x_k_ir)+log(soc_up-x_k_ir))...
                - sum(log(-u_lo(mm)'+u_ir(mm)')+log(u_up(mm)'-u_ir(mm)'));
        Ts=beta*Ts;
    end

    % If decrement is less than threshold, "t" is updated
    if lambda < Epsilon_lambda
        t=mu*t;
        break
    end
end

% Plots
% If watch_evolution == 1, the most important variables are plotted in
% order to verify the evolution of the algorithm

if ii == 1 && watch_evolution == 1
    % Sets up figure(10) to watch Phase I evolution
    figure(10)
    clf(figure(10),'reset')
    set(gcf,'color','w');
```

```
        fig = gcf;
        fig.PaperPositionMode = 'auto';

        set(gcf,'NextPlot','add');
        axes;
        h = title('Phase I: Press any key to continue');
        set(gca,'Visible','off');
        set(h,'Visible','on', 'FontSize',13);

        pause
    end

    if watch_evolution == 1 && sum(ii == [1 0:5:Max_Iter])==1

        x = cumsum([soc_0 u*dt],2);
        x = x(:,2:end);
        TotalLoad = Lhat_+sum(u,1);
        voltages_total = (v_k_base+(A_ev*u)*1000/v_0)';
        power_lim_up = (mask.*u_up)';
        power_lim_lo = (mask.*u_lo)';


        figure(10)
        % State of Charge plots
        subplot(2,2,1);
        plot1 = plot(0:length(Lhat_),[soc_0 x],'LineWidth',1.5);
        for ll=1:N
            set(plot1(ll),'DisplayName',['EV ' num2str(ll)]);
        end
        hold on
        set(gca,'ColorOrderIndex',1);
        plot(0:length(Lhat_),[soc_0 soc_min],':','LineWidth',1.5,...
            'HandleVisibility','off')
        set(gca,'ColorOrderIndex',1);
        plot(0:length(Lhat_),[soc_0 soc_max],':','LineWidth',1.5,...
            'HandleVisibility','off')
        hold off
        set(gca,'YLim',[-2 max(soc_c)+4])
        set(gca,'YTick',0: 2: max(soc_c)+3)
        xlabel(gca,['Time [',num2str(dt),'h]'])
        ylabel(gca,'SoC [kWh]')
        set(gca,'FontSize',12)
        grid
        legend1 = legend(gca,'show');
        set(legend1,'Location','north','FontSize',10,'Orientation',...
            'Horizontal');

        % Total load plots
        subplot(2,2,2);
        plot(0:length(Lhat_)-1,Lhat_,'LineWidth',1.5, 'DisplayName',...
            'Base load')
        hold on
        plot(0:length(Lhat_)-1,TotalLoad,'LineWidth',1.5,'DisplayName',...
            '+ total EV load')
        hold off
        xlabel(gca,['Time [',num2str(dt),'h]'])
        ylabel(gca,'Power [kW]')
        set(gca,'FontSize',12)
        grid
```

```
        legend1 = legend(gca,'show');
        set(legend1,'Location','southwest','FontSize',10);

        % Voltage plots
        subplot(2,2,3);
        plot(0:length(Lhat_)-1,voltages_total,'LineWidth',1.5,...
            'DisplayName','Voltages')
        hold on
        plot(0:length(Lhat_)-1,v_max(1,:)','--','Color',[.31 .31 .31])
        plot(0:length(Lhat_)-1,v_min(1,:)','--','Color',[.31 .31 .31])
        plot(0:length(Lhat_)-1,v_max(1,:)'+s*1e3/v_0,':',...
            'Color',[.31 .31 .31],'LineWidth',2)
        plot(0:length(Lhat_)-1,v_min(1,:)'-s*1e3/v_0,':',...
            'Color',[.31 .31 .31],'LineWidth',2)
        hold off
        set(gca,'YLim',[min(v_min(1,:)-s*1e3/v_0) ...
            max(v_max(1,:)+s*1e3/v_0)])
        xlabel(gca,['Time [',num2str(dt),'h]'])
        ylabel(gca,'Voltage per node [V]')
        set(gca,'FontSize',12)
        grid

        % Charging/discharging rate plots
        subplot(2,2,4);
        plot1 = plot(0:length(Lhat_)-1,u','LineWidth',1.5);
        for ll=1:N
            set(plot1(ll),'DisplayName',['EV ' num2str(ll)]);
        end
        hold on
        set(gca,'ColorOrderIndex',1);
        plot(0:length(Lhat_)-1,power_lim_up,':','LineWidth',1.5,...
            'HandleVisibility','off')
        set(gca,'ColorOrderIndex',1);
        plot(0:length(Lhat_)-1,power_lim_lo,':','LineWidth',1.5,...
            'HandleVisibility','off')
        hold off
        set(gca,'YLim',[-max(p_max_c)-0.5 max(p_max_c)+0.5])
        xlabel(gca,['Time [',num2str(dt),'h]'])
        ylabel(gca,'Power [kW]')
        set(gca,'FontSize',12)
        legend1 = legend(gca,'show');
        set(legend1,'Location','east','FontSize',8);
        grid

        set(gcf,'NextPlot','add');
        axes;
        h = title(['Phase I: Iteration: ', num2str(ii), ',   s: ',...
            num2str(s) ',    lamda: ' num2str(lambda/2) ',   Ts: '...
            num2str(Ts) ',   m/t: ' num2str(num_constraints/t)]);
        set(gca,'Visible','off');
        set(h,'Visible','on', 'FontSize',13);

        pause(0.01)
    end
% Descent computation
u_pre = u;
u_pre(mm) = u(mm) + Ts*Delta_x(1:end-1)';
s_pre = s + Ts*Delta_x(end);
```

```
    % Stooping criteria (Stop specially if "s" goes below zero which means
    % that voltage contraints are feasible)
    if  s < 0 || num_constraints/t < Epsilon_Barrier % stopping criterion
        break
    else
        u = u_pre;
        s = s_pre;
    end
end

% Plot of the final iteration
if watch_evolution == 1
    x = cumsum([soc_0 u*dt],2);
    x = x(:,2:end);
    TotalLoad = Lhat_+sum(u,1);
    voltages_total = (v_k_base+(A_ev*u)*1000/v_0)';
    power_lim_up = (mask.*u_up)';
    power_lim_lo = (mask.*u_lo)';


    figure(10)
    % State of Charge plots
    subplot(2,2,1);
    plot1 = plot(0:length(Lhat_),[soc_0 x],'LineWidth',1.5);
    for ll=1:N
        set(plot1(ll),'DisplayName',['EV ' num2str(ll)]);
    end
    hold on
    set(gca,'ColorOrderIndex',1);
    plot(0:length(Lhat_),[soc_0 soc_min],':','LineWidth',1.5,...
        'HandleVisibility','off')
    set(gca,'ColorOrderIndex',1);
    plot(0:length(Lhat_),[soc_0 soc_max],':','LineWidth',1.5,...
        'HandleVisibility','off')
    hold off
    set(gca,'YLim',[-2 max(soc_c)+4])
    set(gca,'YTick',0: 2: max(soc_c)+3)
    xlabel(gca,['Time [',num2str(dt),'h]'])
    ylabel(gca,'SoC [kWh]')
    set(gca,'FontSize',12)
    grid
    legend1 = legend(gca,'show');
    set(legend1,'Location','north','FontSize',10,'Orientation',...
        'Horizontal');

    % Total load plots
    subplot(2,2,2);
    plot(0:length(Lhat_)-1,Lhat_,'LineWidth',1.5, 'DisplayName',...
        'Base load')
    hold on
    plot(0:length(Lhat_)-1,TotalLoad,'LineWidth',1.5,'DisplayName',...
        '+ total EV load')
    hold off
    xlabel(gca,['Time [',num2str(dt),'h]'])
    ylabel(gca,'Power [kW]')
    set(gca,'FontSize',12)
    grid
    legend1 = legend(gca,'show');
    set(legend1,'Location','southwest','FontSize',10);
```

```
% Voltage plots
subplot(2,2,3);
plot(0:length(Lhat_)-1,voltages_total,'LineWidth',1.5,...
    'DisplayName','Voltages')
hold on
plot(0:length(Lhat_)-1,v_max(1,:)','--','Color',[.31 .31 .31])
plot(0:length(Lhat_)-1,v_min(1,:)','--','Color',[.31 .31 .31])
plot(0:length(Lhat_)-1,v_max(1,:)'+s*1e3/v_0,':',...
    'Color',[.31 .31 .31],'LineWidth',2)
plot(0:length(Lhat_)-1,v_min(1,:)'-s*1e3/v_0,':',...
    'Color',[.31 .31 .31],'LineWidth',2)
hold off
set(gca,'YLim',[min(v_min(1,:)-s*1e3/v_0) max(v_max(1,:)+s*1e3/v_0)])
xlabel(gca,['Time [',num2str(dt),'h]'])
ylabel(gca,'Voltage per node [V]')
set(gca,'FontSize',12)
grid

% Charging/discharging rate plots
subplot(2,2,4);
plot1 = plot(0:length(Lhat_)-1,u','LineWidth',1.5);
for ll=1:N
    set(plot1(ll),'DisplayName',['EV ' num2str(ll)]);
end
hold on
set(gca,'ColorOrderIndex',1);
plot(0:length(Lhat_)-1,power_lim_up,':','LineWidth',1.5,...
    'HandleVisibility','off')
set(gca,'ColorOrderIndex',1);
plot(0:length(Lhat_)-1,power_lim_lo,':','LineWidth',1.5,...
    'HandleVisibility','off')
hold off
set(gca,'YLim',[-max(p_max_c)-0.5 max(p_max_c)+0.5])
xlabel(gca,['Time [',num2str(dt),'h]'])
ylabel(gca,'Power [kW]')
set(gca,'FontSize',12)
legend1 = legend(gca,'show');
set(legend1,'Location','east','FontSize',8);
grid

set(gcf,'NextPlot','add');
axes;
h = title(['Phase I: Iteration: ', num2str(ii), ',   s: ',...
    num2str(s) ',   lamda: ' num2str(lambda/2) ',   Ts: ' ...
    num2str(Ts) ',   m/t: ' num2str(num_constraints/t)]);
set(gca,'Visible','off');
set(h,'Visible','on', 'FontSize',13);

    pause(1)
end

% output assignments
pow_traj_ast = u;

end
```

2.6.3 Phase II Optimisation Routine

This routine optimizes the objective function (cost of charging the PEVs) starting from the initial guesses provided by the Phase I routine.

```
function [pow_traj_ast,soc_traj_ast] = ...
    Book_EV_Centralized_Scheduling_PhaseII(soc_c,soc_d,soc_0,p_max_c,...
    p_max_i,t_0,Ki,dt,A,v_0,v_k_base,v_max,v_min,Lhat,nodes,tariffs,...
    u_0_phaseI,watch_evolution)

% ------ Description inputs
% soc_c : battery available capacities [kWh]
% soc_d : final desired states of charge [kWh]
% soc_0 : initial states of charge [kWh]
% p_max_c : max charging rates [kW]
% p_max_i : max discharging rates [kW]
% t_0 : time step corresponding to soc_0
% Ki : total time steps
% dt : time step
% A : system matrix
% v_0 : transformer voltage
% v_k_base : voltages imposed by the transformer and base loads on the grid
%              nodes
% v_max : upper boundary for voltages on each node
% v_min : lower boundary for voltages on each node
% Lhat_ : total load forecast on the transformer
% nodes : nodes of the grid where EVs are connected
% tariffs : day and night tariffs (or per hour tariffs)
% u_0_phaseI : Initial guess defined by the phase I routine
% watch_evolution : (1) plots internal evolution, (otherwise) ignores

% ------ Description outputs
% pow_traj_ast : optimal power trajectory
% s : max infeasibility bound on the voltage constraints

% ------ Onjective
% Minimize the cost of charging the three PEVs while respecting the power,
% soc and voltage constraints

% Initialization of variables

% number of EVs
N = length(Ki);
Lhat_=Lhat';

% construction of soc_k constraints for visualization of evolution
soc_min = zeros(N,length(Lhat_)+1);
soc_min(sub2ind(size(soc_min),(1:N)',Ki*0+1)) = soc_0-1e-3;
soc_min(sub2ind(size(soc_min),(1:N)',Ki*0+t_0+2)) = -soc_0+2e-3;
soc_min(sub2ind(size(soc_min),(1:N)',Ki+t_0+1)) = +soc_d-2e-3;
soc_min = cumsum(soc_min,2);
soc_min=soc_min(:,2:end);

soc_max = zeros(N,length(Lhat_)+1);
soc_max(sub2ind(size(soc_max),(1:N)',Ki*0+1)) = soc_0+1e-3;
soc_max(sub2ind(size(soc_max),(1:N)',Ki*0+t_0+2)) = -soc_0+soc_c-2e-3;
soc_max(sub2ind(size(soc_max),(1:N)',Ki+t_0+1)) = -soc_c+soc_d+2e-3;
```

```
soc_max = cumsum(soc_max,2);
soc_max=soc_max(:,2:end);

% construction of soc_k constraints for the routine
soc_lo = zeros(N,length(Lhat_)+1);
soc_lo(sub2ind(size(soc_lo),(1:N)',Ki*0+1)) = +1e-3;
soc_lo(sub2ind(size(soc_lo),(1:N)',Ki*0+t_0+2)) = -soc_0-2e-3;
soc_lo(sub2ind(size(soc_lo),(1:N)',Ki+t_0+1)) = Inf;
soc_lo = cumsum(soc_lo,2);
soc_lo=soc_lo(:,2:end);
soc_lo(:,1:t_0)=[];
soc_lo(:,max(Ki):end)=[];
soc_lo=soc_lo(:);
soc_lo(soc_lo==Inf)=[];


soc_up = zeros(N,length(Lhat_)+1);
soc_up(sub2ind(size(soc_up),(1:N)',Ki*0+1)) = -1e-3;
soc_up(sub2ind(size(soc_up),(1:N)',Ki*0+t_0+2)) = -soc_0+soc_c+2e-3;
soc_up(sub2ind(size(soc_up),(1:N)',Ki+t_0+1)) = Inf;
soc_up = cumsum(soc_up,2);
soc_up=soc_up(:,2:end);
soc_up(:,1:t_0)=[];
soc_up(:,max(Ki):end)=[];
soc_up=soc_up(:);
soc_up(soc_up==Inf)=[];

% construction of charging rate constraints for each EV
u_up = zeros(N,length(Lhat_));
u_up(sub2ind(size(u_up),(1:N)',Ki*0+t_0+1*(t_0==0))) = -p_max_c*(t_0>0);
u_up(sub2ind(size(u_up),(1:N)',Ki+t_0)) = p_max_c;
u_up = cumsum(u_up,2,'reverse')+1e-3;


u_lo = zeros(N,length(Lhat_));
u_lo(sub2ind(size(u_lo),(1:N)',Ki*0+t_0+1*(t_0==0))) = +p_max_i*(t_0>0);
u_lo(sub2ind(size(u_lo),(1:N)',Ki+t_0)) = -p_max_i;
u_lo = cumsum(u_lo,2,'reverse')-1e-3;

% construction of initial guess of charging rate trajectories
% note: Inherited from Phase I
u = u_0_phaseI;

% construction of a mask to avoid influence on unexisting controls
mask = zeros(N,length(Lhat_)+1);
mask(sub2ind(size(mask),(1:N)',Ki+t_0+1)) = 1;
mask(sub2ind(size(mask),(1:N)',Ki*0+t_0+1)) = -1;
mask = cumsum(mask,2,'reverse');
mask=mask(:,2:end);
mm = (mask(:)==1)';

% constraints on A_ev*u defined by constraints on voltages of each node
v_up = (v_0/1000)*(v_max(:,sum(mask,1)>0)-v_k_base(:,sum(mask,1)>0))+1e-3;
v_lo = (v_0/1000)*(v_min(:,sum(mask,1)>0)-v_k_base(:,sum(mask,1)>0))-1e-3;

% construction of equality constraints matrix "A_soc_end" for the final
% states of charge
A_soc_end = zeros(N,N*max(Ki));
for tt = 1 : max(Ki)
    A_soc_end(:,(tt-1)*N+1:(tt)*N) = diag(mask(:,t_0+tt))*dt;
end
```

```
% construction of inequality constraints matrix "A_soc" for the partial
% states of charge at each time step "k"
A_soc = repmat(A_soc_end,max(Ki)-1,1);
for nn = 1 : max(Ki)-1
    A_soc((nn-1)*N+1:nn*N,(nn)*N+1:end) = ...
        0*A_soc((nn-1)*N+1:nn*N,(nn)*N+1:end);
end
Indices=zeros(1,sum(max(Ki)-Ki));
prev_count = 0;
for ii = 1:length(Ki)
    Indices_Ki = (Ki(ii)):(max(Ki)-1);
    Indices(prev_count+1:prev_count+length(Indices_Ki)) = ...
        (Indices_Ki-1)*N+ii;
    prev_count = prev_count +length(Indices_Ki);
end
A_soc(Indices,:)=[];
A_soc(:,sum(A_soc_end)==0)=[];
A_soc_end(:,sum(A_soc_end)==0)=[];

% Elimination of nodes (columns of the system matrix A) without PEVs.
A_ev = A(:,nodes);
% Repetition of "A_ev" matrix for each time step until the disconnection of
% the last PEV.
A_ev_re = zeros(length(A)*max(Ki),sum(Ki));
prev_row = 0;
prev_col = 0;
for tt = 1:max(Ki)
    N_connected = sum(mask(:,t_0+tt));
    A_ev_re(prev_row+1:prev_row+length(A),...
        prev_col+1:prev_col+N_connected) = ...
        A_ev(:,mask(:,t_0+tt)==1);
    prev_row = prev_row+length(A);
    prev_col = prev_col+N_connected;
end

% Parameters of descent, and backtracking line search
alpha = 0.5;
beta = 0.8;
mu = 10;
t = 1;
t_pre = t;
num_constraints = 2*sum(Ki)+2*size(A_soc,1)+2*size(A_ev_re,1);
Epsilon_lambda = 1e-20;
Epsilon_Barrier = 1e-50;

% Max number of iterations
Max_Iter=1000;

for ii = 1:Max_Iter

    % computes the states of charge trajectories
    x = cumsum([soc_0 u*dt],2);
    x = x(:,2:end);

    x_k = A_soc*u(mm)';

    % computes voltage contribution of EVs
    v = A_ev_re*u(mm)';
```

```
% computes the extended objective function's current value (including
% inequalities implicitly)
H = t*sum(tariffs*u') ...
          - sum(log(-soc_lo+x_k)+log(soc_up-x_k))...
          - sum(log(-u_lo(mm)'+u(mm)')+log(u_up(mm)'-u(mm)'))...
          - sum(log(-v_lo(:)+v)+log(v_up(:)-v));

% computes the gradient of the extended objective function
dH_du0 = repmat(tariffs,N,1);

dH_du = t*dH_du0(mm)' ...
          + A_soc'*(1./(soc_lo-x_k)+1./(soc_up-x_k))...
          + eye(sum(Ki))'*(1./(u_lo(mm)'-u(mm)')+...
                           1./(u_up(mm)'-u(mm)'))...
          + A_ev_re'*(1./(v_lo(:)-v)+1./(v_up(:)-v));

% computes de Hessian given the implicit inclusion of the inequalities
% in the objective function
I_x_lo = diag(1./(soc_lo-x_k).^2);
I_x_up = diag(1./(soc_up-x_k).^2);

I_u_lo = diag(1./(u_lo(mm)'-u(mm)').^2);
I_u_up = diag(1./(u_up(mm)'-u(mm)').^2);

I_v_lo = diag(1./(v_lo(:)-v).^2);
I_v_up = diag(1./(v_up(:)-v).^2);

Hessian = A_ev_re'*I_v_lo*A_ev_re + A_ev_re'*I_v_up*A_ev_re +...
          I_u_lo + I_u_up + A_soc'*I_x_lo*A_soc + A_soc'*I_x_up*A_soc;


% normalizaion of the gradient
dH_du_normalized = dH_du/norm(dH_du);

% computes the direction of descent in the Newton's method with
% equality constraints
if rcond([Hessian A_soc_end';A_soc_end zeros(size(A_soc_end,1))])>1e-15
    Delta_x_w = ...
          -[Hessian A_soc_end';A_soc_end zeros(size(A_soc_end,1))]\...
          [dH_du_normalized;zeros(size(A_soc_end,1),1)];
else % If the extended matrix in the equality constrained
     % problem is badly conditionned, the routine stops
    break
end

% Extracts the elements that correspond to the direction of descent
Delta_x = Delta_x_w(1:length(dH_du_normalized));
% w = Delta_x_w(length(dH_du_normalized)+1:end);

% Computes the decrement
lambda = (Delta_x'*Hessian*Delta_x)^2;


% If the decrement is less that the threshold after changing "t",
% the routine stops
if t > t_pre && lambda < Epsilon_lambda
    break
else
```

```
end
t_pre = t;

% Initial step size in the backtracking line search
Ts = 100000;

H_ir = Inf;

% backtracking line search subroutine
while (H_ir > (H+alpha*Ts*(dH_du_normalized'*Delta_x)))

    u_ir = u;
    u_ir(mm) = u(mm) + Ts*Delta_x';
    % computes the possible state of charge trajectories

    x_k_ir = A_soc*u_ir(mm)';


    % computes possible voltage contribution of EVs
    v_ir = A_ev_re*u_ir(mm)';

    % checks is possible variable values lie within specified ranges
    if (sum(u_ir(mm)' < u_lo(mm)') + sum(u_ir(mm)' > u_up(mm)') ...
            + sum(x_k_ir < soc_lo) + sum(x_k_ir > soc_up) ...
            + sum(v_ir < v_lo(:)) + sum(v_ir > v_up(:)))>0
        Ts=beta*Ts;
    else
        H_ir = t*tariffs*sum(u_ir,1)' ...
            - sum(log(-soc_lo+x_k_ir)+log(soc_up-x_k_ir))...
            - sum(log(-u_lo(mm)'+u_ir(mm)')+log(u_up(mm)'-u_ir(mm)'))...
            - sum(log(-v_lo(:)+v_ir)+log(v_up(:)-v_ir));
        Ts=beta*Ts;
    end

    % If decrement is less than threshold, "t" is updated
    if lambda < Epsilon_lambda
        t=mu*t;
        break
    end
end

% Plots
% If watch_evolution == 1, the most important variables are plotted in
%  order to verify the evolution of the algorithm

if ii == 1 && watch_evolution == 1
    % Sets up figure(11) to watch Phase II evolution
    figure(11)
    clf(figure(11),'reset')
    set(gcf,'color','w');
    fig = gcf;
    fig.PaperPositionMode = 'auto';

    set(gcf,'NextPlot','add');
    axes;
    h = title('Phase II: Press any key to continue');
    set(gca,'Visible','off');
    set(h,'Visible','on', 'FontSize',13);
```

```
        pause
end

if watch_evolution == 1 && sum(ii == [1 0:5:Max_Iter])==1
    figure(11)
    % State of Charge plots
    subplot(2,2,1);
    plot1 = plot(0:length(Lhat_),[soc_0 x],'LineWidth',1.5);
    for ll=1:N
        set(plot1(ll),'DisplayName',['EV ' num2str(ll)]);
    end
    hold on
    set(gca,'ColorOrderIndex',1);
    plot(0:length(Lhat_),[soc_0 soc_min],':','LineWidth',1.5,...
        'HandleVisibility','off')
    set(gca,'ColorOrderIndex',1);
    plot(0:length(Lhat_),[soc_0 soc_max],':','LineWidth',1.5,...
        'HandleVisibility','off')
    hold off
    set(gca,'YLim',[-2 max(soc_c)+4])
    set(gca,'YTick',0: 2: max(soc_c)+3)
    xlabel(gca,['Time [',num2str(dt),'h]'])
    ylabel(gca,'SoC [kWh]')
    set(gca,'FontSize',12)
    grid
    legend1 = legend(gca,'show');
    set(legend1,'Location','north','FontSize',10,'Orientation',...
        'Horizontal');

    % Total load plots
    subplot(2,2,2);
    plot(0:length(Lhat_)-1,Lhat_,'LineWidth',1.5, 'DisplayName',...
        'Base load')
    hold on
    plot(0:length(Lhat_)-1,Lhat_+sum(u,1),'LineWidth',1.5,...
        'DisplayName','+ total EV load')
    hold off
    xlabel(gca,['Time [',num2str(dt),'h]'])
    ylabel(gca,'Power [kW]')
    set(gca,'FontSize',12)
    grid
    legend1 = legend(gca,'show');
    set(legend1,'Location','southwest','FontSize',10);

    % Voltage plots
    subplot(2,2,3);
    plot(0:length(Lhat_)-1,(v_k_base+(A_ev*u)*1000/v_0)',...
        'LineWidth',1.5, 'DisplayName',...
        'Voltages')
    hold on
    plot(0:length(Lhat_)-1,v_max(1,:)','--','Color',[.31 .31 .31])
    plot(0:length(Lhat_)-1,v_min(1,:)','--','Color',[.31 .31 .31])
    hold off
    xlabel(gca,['Time [',num2str(dt),'h]'])
    ylabel(gca,'Voltage per node [V]')
    set(gca,'FontSize',12)
    grid

    % Charging/discharging rate plots
```

```
    subplot(2,2,4);
    plot1 = plot(0:length(Lhat_)-1,u','LineWidth',1.5);
    for ll=1:N
        set(plot1(ll),'DisplayName',['EV ' num2str(ll)]);
    end
    hold on
    set(gca,'ColorOrderIndex',1);
    plot(0:length(Lhat_)-1,(mask.*u_up)',':','LineWidth',1.5,...
        'HandleVisibility','off')
    set(gca,'ColorOrderIndex',1);
    plot(0:length(Lhat_)-1,(mask.*u_lo)',':','LineWidth',1.5,...
        'HandleVisibility','off')
    hold off
    set(gca,'YLim',[-max(p_max_c)-0.5 max(p_max_c)+0.5])
    xlabel(gca,['Time [',num2str(dt),'h]'])
    ylabel(gca,'Power [kW]')
    set(gca,'FontSize',12)
    legend1 = legend(gca,'show');
    set(legend1,'Location','east','FontSize',8);
    grid

    set(gcf,'NextPlot','add');
    axes;
    h = title(['Phase II: Iteration: ' num2str(ii) ',   lamda: '...
        num2str(lambda/2) ',   Ts: ' num2str(Ts) ',   m/t: '...
        num2str(num_constraints/t)]);
    set(gca,'Visible','off');
    set(h,'Visible','on',  'FontSize',13);

    pause(0.01)

end

% Descent computation
u_pre = u;
u_pre(mm) = u(mm) + Ts*Delta_x';

% Stooping criteria
if num_constraints/t<Epsilon_Barrier % stopping criterion
    break
else
    u = u_pre;
end
end
% Plot of the final iteration
if watch_evolution == 1

figure(11)
% State of Charge plots
subplot(2,2,1);
plot1 = plot(0:length(Lhat_),[soc_0 x],'LineWidth',1.5);
for ll=1:N
    set(plot1(ll),'DisplayName',['EV ' num2str(ll)]);
end
hold on
set(gca,'ColorOrderIndex',1);
plot(0:length(Lhat_),[soc_0 soc_min],':','LineWidth',1.5,...
    'HandleVisibility','off')
set(gca,'ColorOrderIndex',1);
```

```matlab
plot(0:length(Lhat_),[soc_0 soc_max],':','LineWidth',1.5,...
    'HandleVisibility','off')
hold off
set(gca,'YLim',[-2 max(soc_c)+4])
set(gca,'YTick',0: 2: max(soc_c)+3)
xlabel(gca,['Time [',num2str(dt),'h]'])
ylabel(gca,'SoC [kWh]')
set(gca,'FontSize',12)
grid
legend1 = legend(gca,'show');
set(legend1,'Location','north','FontSize',10,'Orientation',...
    'Horizontal');

% Total load plots
subplot(2,2,2);
plot(0:length(Lhat_)-1,Lhat_,'LineWidth',1.5, 'DisplayName',...
    'Base load')
hold on
plot(0:length(Lhat_)-1,Lhat_+sum(u,1),'LineWidth',1.5,...
    'DisplayName','+ total EV load')
hold off
xlabel(gca,['Time [',num2str(dt),'h]'])
ylabel(gca,'Power [kW]')
set(gca,'FontSize',12)
grid
legend1 = legend(gca,'show');
set(legend1,'Location','southwest','FontSize',10);

% Voltage plots
subplot(2,2,3);
plot(0:length(Lhat_)-1,(v_k_base+(A_ev*u)*1000/v_0)',...
    'LineWidth',1.5, 'DisplayName','Voltages')
hold on
plot(0:length(Lhat_)-1,v_max(1,:)','--','Color',[.31 .31 .31])
plot(0:length(Lhat_)-1,v_min(1,:)','--','Color',[.31 .31 .31])
hold off
xlabel(gca,['Time [',num2str(dt),'h]'])
ylabel(gca,'Voltage per node [V]')
set(gca,'FontSize',12)
grid

% Charging/discharging rate plots
subplot(2,2,4);
plot1 = plot(0:length(Lhat_)-1,u','LineWidth',1.5);
for ll=1:N
    set(plot1(ll),'DisplayName',['EV ' num2str(ll)]);
end
hold on
set(gca,'ColorOrderIndex',1);
plot(0:length(Lhat_)-1,(mask.*u_up)',':','LineWidth',1.5,...
    'HandleVisibility','off')
set(gca,'ColorOrderIndex',1);
plot(0:length(Lhat_)-1,(mask.*u_lo)',':','LineWidth',1.5,...
    'HandleVisibility','off')
hold off

set(gca,'YLim',[-max(p_max_c)-0.5 max(p_max_c)+0.5])
xlabel(gca,['Time [',num2str(dt),'h]'])
ylabel(gca,'Power [kW]')
```

```
        set(gca,'FontSize',12)
        legend1 = legend(gca,'show');
        set(legend1,'Location','east','FontSize',8);
        grid

        set(gcf,'NextPlot','add');
        axes;
        h = title(['Phase II: Iteration: ' num2str(ii) ',    lamda: '...
            num2str(lambda/2) ',    Ts: '  num2str(Ts) ',    m/t: '...
            num2str(num_constraints/t)]);
        set(gca,'Visible','off');
        set(h,'Visible','on',  'FontSize',13);

        pause(1)

    end
    % output assignments
    pow_traj_ast = u;
    soc_traj_ast = cumsum([soc_0 u*dt],2);
    end
```

2.7   Conclusion

This chapter provides a linear approach to compute voltages on a low voltage grid topology based on the house instantaneous load and the instantaneous consumption/injection of PEVs. Based on this linear modeling, the chapter proposes a centralized formulation to manage the charging of multiple PEVs on a residential grid under certain assumptions. Additionally, the proposed strategy employs the energy storage capacity of the PEVs to provide a voltage support services to the grid. The proposed approach is tested under multiple scenarios to test its ability to maintain voltages within limits and provide optimal consumption/injection policies.

In terms of perspectives for these type of centralized approaches, multiple alternative tests can be considered. For instance, the inclusion of a differentiation between energy selling prices and buying prices, and the inclusion of other controllable home appliances in the optimal scheduling. However, multiple drawbacks become evident with centralized schedulers. The first problem is the centralization of information from PEV owners. The scheduler requires the information of connected PEVs to run the scheduling routine. Furthermore, the model of the grid has as inputs the forecasts at each bus where a load is connected. This demands pre-processing and memory requirements for each of the concerned buses.

Focusing on the approach of this chapter and the evaluated scenarios, several degrees of freedom can be adjusted to take advantage of bidirectional chargers. It has been observed that differential tariffs motivate PEV energy needs to be drawn from the grid at convenient hours. Moreover, it has been observed that voltage constraints can be tuned to avoid undesirable effects on the total load. Nevertheless, it can be observed as well, that the effect of voltage constraints is directly observed on the total load distribution over time. This leads to the conclusion that the load scheduling efforts can be focused on managing this total load behavior with the PEVs load scheduling, instead of managing the behavior of the voltages of the grid. In other words it can be observed that if PEV energy consumption/injection is managed such that the total load profiles during the day are as uniform as possible, then the effect on voltage levels will be positive. As it has been observed in the last section, peaks of consumption are reflected on voltages sags. Thus, uniform load profiles over time result in uniform voltage level profiles over time. Then, instead of handling a wide set of inputs for a grid model, it is much more feasible and practical to handle total load forecasts as inputs of these PEV load management approaches.

Chapter 3
Dynamic Programming and Potential Game Approach

In this chapter, a decentralized PEV load scheduling approach is described. It combines a forward induction dynamic programming (DP) algorithm with a potential game (game theory) framework. The proposed approach optimally manages PEV charging while it handles different constraints imposed by the charger, the state of charge, and the charging duration. A detailed explanation of the DP algorithm adapted for a single PEV is given. It is followed by a description of the N-person non-cooperative potential game framework for the extension of the scheduling approach to multiple PEVs. The potential function in this potential game is intended to reduce the distance between the transformer's total load and its average over a time horizon. Thus, PEV energy storage capacities are used for compensating load peaks and shifting consumption to valley hours. Several simple case studies are analyzed. To compare with the centralized method of Chap. 2, the proposed decentralized approach is also tested under more realistic conditions using the IEEE European low voltage test feeder. The chapter ends with a Matlab script for the forward dynamic programming algorithm. Readers are encouraged to use this script jointly with scripts of the following chapters to test the example results and improve performances.

3.1 Introduction

This chapter discusses the details of a decentralized approach to optimally manage PEV charging schedules based on Dynamic Programming (DP) and Game Theory [OHB15]. The chapter provides a detailed explanation of a forward induction DP algorithm and its adaptation to the problem of optimal charging of a single PEV with its corresponding constraints on power consumption, upper and lower limits on states of charge, desired states of charge, etc. Given the fact that the state and control action spaces are finite, the optimal policies can be tractably computed with the induction algorithm [WOB14]. The extension to multiple PEVs is provided by the adaptation of a N-person non-cooperative potential game. In this game, the payoff of

© Springer International Publishing AG 2018
A. Ovalle et al., *Grid Optimal Integration of Electric Vehicles: Examples with Matlab Implementation*, Studies in Systems, Decision and Control 137, https://doi.org/10.1007/978-3-319-73177-3_3

each player is based on a utility function that aims to reduce the distance between the total load and the average load, achieving load profile flattening. The concept behind the strategies for each player in the game is detailed to associate it with the dynamic programming algorithm. Several study case are employed to test the approach and compare it with the centralized approach of Chap. 2.

3.2 Optimal Charging Problem

Recalling from Chap. 2, the problem of charging multiple plug-in electric vehicles (PEV) subject to a certain number of constraints can be addressed as follows,

$$\min_{\mathbf{x}^i \in \Theta^i, i=\{1,\cdots,N\}} cost(\mathbf{x}^1, \mathbf{x}^2, \cdots, \mathbf{x}^i, \cdots, \mathbf{x}^J) \tag{3.1}$$

where $cost(\cdot)$ is an objective function having as parameters the power consumption schedules $\mathbf{x}^i$ of the N electric vehicles to be charged. The power consumption/injection variables at each instant of a discretized charging period are the elements of the vector $\mathbf{x}^i$,

$$\mathbf{x}^i = \left[x_1^i, x_2^i, \cdots, x_k^i, \cdots, x_{K^i}^i \right].$$

The set Θ^i of all possible power consumption schedules for the vehicle i can be defined by the constraints described as follows. The *power consumption/injection* limits depend on the type of charger. These limits can be expressed by the following inequalities,

$$-\overline{p}^i \le x_k^i \le \overline{p}^i, \quad \forall k \in \left\{1, \cdots, K^i\right\}, \forall i \in \{1, \cdots, J\},$$

where $\overline{p}^i$ is the nominal power of the charger of PEV i. The *partial states of charge* denoted by soc_k^i must always be bounded by a certain upper limit $(\overline{soc}^i)$ and a lower limit $(\underline{soc}^i)$ as it is expressed by,

$$\underline{soc}^i \le soc_0^i + \tau \sum_{\kappa=1}^{k} x_\kappa^i \le \overline{soc}^i, \quad \forall k \in \left\{1, \cdots, K^i - 1\right\}, \forall i \in \{1, \cdots, J\},$$

where τ is the length of time steps. Both the upper and lower limits are defined to reduce the impact of charging and discharging cycles on batteries life spans. The final or *desired states of charge* soc_d^i can be defined by equality constraints as follows,

$$soc_0^i + \tau \sum_{k=1}^{K^i} x_k^i = soc_d^i, \quad \forall i \in \{1, \cdots, J\},$$

or as an intervals of desirable states of charge (more relaxed constraints) as follows,

$$soc_d^i \leq soc_i^0 + \tau \sum_{k=1}^{K^i} x_k^i \leq \overline{soc}^i, \quad \forall i \in \{1, \cdots, J\}.$$

Additionally, other types of constraints can be included. For example, with an approximated model of the grid, the voltage level at every node of the grid can be included as a constraint to provide a voltage support service with the storage capacity of the batteries as it was introduced on Chap. 2 [Ova+14, CNHD10, CNHD11]. However, the inherent centralization of this type of approaches demands great computation capacity, and collecting and handling great quantities of information. This chapter provide details of a forward induction dynamic programming algorithm adapted to solve the charging schedule problem of one PEV. Then, a non-cooperative N-person game is presented to link the DP solvers of multiple PEVs and optimally manage the charging of multiple PEVs in a decentralized scheme.

3.3 Dynamic Programming Algorithm

The dynamic programming technique is usually employed to solve certain types of problems where decisions have to be taken in stages. In this type of situations, the objective is to take a sequence of decisions to minimize an undesirable outcome or maximize a profit. Usually, this type of problems can be represented as discrete-time dynamic systems where the outcome is an additive function over time (discrete stages). The sequence of decisions are taken to influence the state variables that describe the system [Rif+11, LD08, Ber12, Bil+14, Hae+14]. The system can be described as follows,

$$\chi_{k+1} = f_k(\chi_k, u_k, \xi_k), \quad k = \{0, 1, \cdots, K-1\}, \tag{3.2}$$

where k is the discrete stage index, f_k is the function that describes the system, χ_k is system's state at time instant k, u_k is the control decision taken in k to steer the system from χ_k to χ_{k+1}, and ξ_k is a random variable that depends on the type of system.

The expected value, with respect to the joint distribution of the random variables, of the total outcome function is expressed as,

$$\mathbb{E}\left\{ \sum_{k=0}^{K-1} g_k(\chi_k, u_k, \xi_k) + g_K(\chi_K) \right\}$$

where $g_k(\chi_k, u_k, \xi_k)$ is the cost at time k, which depends on the of state, the control, and the random variables involved. Let us define *admissible control laws* as sequences

$\Lambda = \{\delta_0, \delta_1, \cdots, \delta_{K-1}\}$ of functions $\delta_k(\chi_k) = u_k$ mapping states χ_k into controls u_k, such that $u_k \in U_k(\chi_k) \subset \mathbb{U}_k$. The set $\mathbb{U}_k$ is the set of all possible control decisions at time k and $U_k(x_k)$ is a non-empty subset of $\mathbb{U}_k$ that depends on the current state $\chi_k \in \Upsilon_k$. The set Υ_k contains all possible states χ_k at time k.

Given an initial state χ_0 and an admissible control law $\Lambda = \{\delta_0, \delta_1, \cdots, \delta_{K-1}\}$, the total expected cost associated to this control law, starting at state χ_0, is defined as,

$$J_\Lambda(\chi_0) = \mathbb{E}\left\{\sum_{k=0}^{K-1} g_k(\chi_k, \delta_k(\chi_k), \xi_k) + g_K(\chi_K)\right\}.$$

It can be noticed that the expected cost of the admissible control law depends on the initial state χ_0. Now, knowing the set $\mathbb{L}$ of admissible control laws, an *optimal control law* Λ^*, starting from χ_0, can be found as,

$$J_{\Lambda^*}(\chi_0) = J^*(\chi_0) = \min_{\Lambda \in \mathbb{L}} J_\Lambda(\chi_0). \tag{3.3}$$

The optimal control law is the optimal sequence of decisions. The function $J^*(\chi_0)$ is also called *optimal cost function*.

3.3.1 Finite-State Problem and the DP Algorithm

Let us consider the case where the state variable at each time k is discretized with a defined step size. Under this conditions, the problem looks like a a multi-stage shortest path problem as the one shown on Fig. 3.1a.

Neglecting the presence of random noise, let us assume that applying feasible control actions u_{k-1}, the cost at time $k - 1$ depends only on the transition between states χ_{k-1} and χ_k, and it is given by a function $g_k(\chi_{k-1}, \chi_k)$. If a control action being able to bring the system from state χ_{k-1} to χ_k is unfeasible, it is represented by a cost function $g_k(\chi_{k-1}, \chi_k) = \infty$. On Fig. 3.1a, if a feasible control exists, the transition is represented by an arc and the cost of transition is given by $g_k(\chi_{k-1}, \chi_k)$. With this assumptions, the Dynamic Programming Algorithm is presented as the pseudo-code of Algorithm 1 [Rif+11].

The general idea is to solve the optimization problem splitting the optimal cost function (3.3) in temporal sub-functions of optimal cost $J_k(\chi_k)$, based on the Bellman's *principle of optimality* [Ber12]. The algorithm finds the optimal trajectory of the state variable and the optimal control sequence in an iterative scheme over the time. The approach followed in this chapter is the forward induction which is illustrated in Fig. 3.1b. The algorithm proceeds forward in time (with the first loop) from $k = 2$ until the final state in $k = K$. For each possible state $\chi_k \in \Upsilon_k$, the algorithm finds the path of optimal cost starting from the root state at stage 0 and leading to the evaluated state in χ_k. For instance, on Fig. 3.1b the algorithm is in the process of constructing the sub-function of optimal cost $J_k(\chi_k)$ knowing that the sub-function of

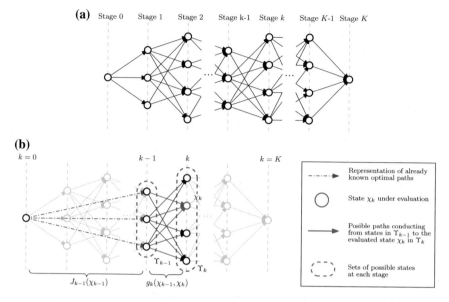

Fig. 3.1 Multi-stage shortest path problem: **a** Multi-stage problem example, **b** Illustration of the Forward Dynamic Programming algorithm. © [2017] IEEE. Reprinted, with permission, from [OHB15]

Algorithm 1 Forward Dynamic Programming Algorithm

1: **for all** $\chi_1 \in \Upsilon_1$ **do**
2:     $J_1(\chi_1) = g_1(x_0, x_1)$
3: **end for**
4: **for** $k = \{2, 3, \cdots, K\}$ **do**
5:     **for all** $\chi_k \in \Upsilon_k$ **do**
6:

$$J_k(\chi_k) = \min_{\chi_{k-1} \in \Upsilon_{k-1}} \left[g_k(\chi_{k-1}, \chi_k) + J_{k-1}(\chi_{k-1}) \right]$$

7:     **end for**
8: **end for**

---

optimal cost $J_{k-1}(\chi_{k-1})$ has already been constructed. In other words, the algorithm is constructing, and keeping in memory, the optimal paths (and their associated costs) to go from χ_0 to all possible states $\chi_k \in \Upsilon_k$, given that it already knows the optimal paths to go from χ_0 to all possible states $\chi_{k-1} \in \Upsilon_{k-1}$, represented in Fig. 3.1b with direct blue arrows. Repeating this procedure until the final stage, the *optimal cost function* $J^*(\chi_0)$ is constructed.

3.3.2 The PEV Battery Charging Problem Adapted to the DP Algorithm

This section describes one possible adaptation of the DP algorithm for the optimal battery charging problem of a PEV. The deterministic version of the system can be described as follows. The state of charge at the end of time step k can be chosen as the state variable χ_k of the system. Thus, the expression describing the system (i.e. Eq. (3.2)) is given by,

$$soc_k^i = soc_{k-1}^i + \tau x_k^i,$$

where the soc_k^i is the state of charge of vehicle i at the end of time step k, and x_k^i is the power consumption rate during the time step. As it can be inferred, the control variable here is the power consumption rate x_k^i, and the set $\mathbb{U}_k^i$ of possible control decisions is defined by,

$$-\overline{p}^i \le x_k^i \le \overline{p}^i, \quad \forall k \in \{0, \cdots, K^i - 1\}.$$

The set Υ_k^i of possible states of the state variable (for PEV i at time k) is defined, in this case, depending on the limits that must be imposed to the state of charge at each time step k. A priori, the boundaries are defined by the lower and upper constraints, $\underline{soc}^i$ and $\overline{soc}^i$. However, some other boundaries are fixed depending on the initial state of charge soc_0^i, on the minimal desired state of charge soc_d^i at the end of the charging period, and on the current state soc_k^i. These boundaries are represented by colored dashed lines (blue and red) on Fig. 3.2a, and are defined by,

$$\underline{soc}_k^i \le soc_k^i \le \overline{soc}_k^i,$$

where $\underline{soc}_k$ and $\overline{soc}_k$ are functions that depend on k as follows,

$$\underline{soc}_k^i = \max \left\{ \underline{soc}^i, \; soc_0^i - (\tau \overline{p}^i)k, \; soc_d^i + (\tau \overline{p}^i)\left(k - K^i\right) \right\}, \qquad (3.4)$$

$$\overline{soc}_k^i = \min \left\{ \overline{soc}^i, \; soc_0^i + (\tau \overline{p}^i)k \right\}. \qquad (3.5)$$

As it can be noticed on Fig. 3.2a, $\underline{soc}_k^i$ and $\overline{soc}_k^i$ depend not only on the upper and lower limits allowed for the state of charge, but also on the maximum allowed rate of charge/discharge (the maximum input/output power of the charger $\overline{p}^i$), and the time step length τ (in hours or fractions of hour).

On the other hand, the subset $U_k^i \subset \mathbb{U}_k^i$ of feasible control decisions, for this case, depends on the current state of the state variable soc_k^i in the sense that at certain states of charge, certain rates of charge/discharge x_k^i are forbidden. For example, if the state of charge at time k is close enough to the maximal state of charge $\overline{soc}^i$, then it is not possible to charge the battery at rates x_k^i producing future states of charge over the limit, i.e. $soc_{k+1} > \overline{soc}^i$. In this sense, and for the purpose of the forward DP algorithm adaptation, knowing the arrival state of charge soc_k^i, the feasible root

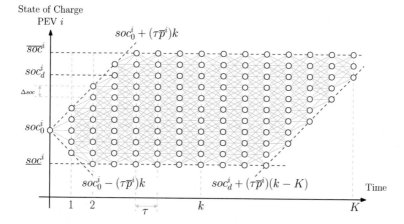

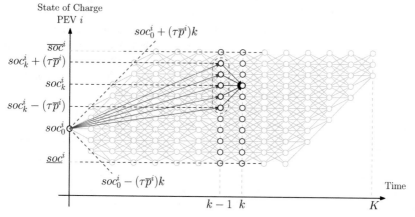

Fig. 3.2 Adaptation of the problem of one PEV charging schedule to the Forward Dynamic Programming Algorithm. © [2017] IEEE. Reprinted, with permission, from [OHB15]

states in $k - 1$ leading to soc_k^i by applying feasible control decisions, are defined by,

$$\underline{soc}_{k-1}^i \leq soc_{k-1}^i \leq \overline{soc}_{k-1}^i$$

where $\underline{soc}_{k-1}^i$ and $\overline{soc}_{k-1}^i$ are functions similar to Eqs. (3.4) and (3.5), but depending on soc_k^i as follows,

$$\underline{soc}_{k-1}^i = \max\left\{ \underline{soc}^i,\ soc_0^i - (\tau\overline{p}^i)(k - 1),\ soc_k^i - \tau\overline{p}^i \right\}$$

$$\overline{soc}_{k-1}^i = \min\left\{ \overline{soc}^i,\ soc_0^i + (\tau\overline{p}^i)(k - 1),\ soc_k^i + \tau\overline{p}^i \right\}.$$

These limits are better illustrated on Fig. 3.2b. Based on these inequalities, the subset $U_k^i(soc_k^i)$ is defined by,

$$\frac{soc_k^i - \overline{soc}_{k-1}^i}{\tau} \leq x_k^i \leq \frac{soc_k^i - \underline{soc}_{k-1}^i}{\tau}$$

Figure 3.3 allows to have a wider view on the evolution of the FDP algorithm for one PEV. In this figure, the whole procedure is illustrated, from the discretization of the state space, until obtaining the final optimal trajectory of the state variable. Once the state space has been discretized both in terms of time and energy, the constraints of Sect. 3.2 are considered, as it was explained with Fig. 3.2. After this, the set of feasible trajectories of the state of charge variable are defined, and the algorithm is able to start the exploration and the construction of the *optimal cost function*. As it was described, the FDP algorithm starts at $k = 2$, and explores all the possible paths from the initial state of charge soc_0^i, to all the possible states of charge at time step $k = 2$. The algorithm stores all the optimal paths for this stage, and their associated costs. Thus, once the algorithm starts the same procedure with all the possible states at $k = 3$, the optimal paths for states in the preceding time step ($k = 2$) will be already stored on memory. The algorithm continues with this procedure until the end of the horizon time $k = K$ when the final optimal path is obtained.

It is possible to observe that memory demand increases as the algorithm advances through the time horizon. Consequently, if the discretization steps (on both time and energy) are refined, then both computational time and memory demands will increase. Furthermore, including more than one PEV in the DP algorithm will exponentially increase the quantity of viable states to visit at each step of time. For instance, if 3 PEVs are considered with exactly the same conditions like the one on the example of Fig. 3.3, then at $k = 1$ there will be 125 combinations of viable states to visit (5 feasible states per PEV), and at $k = 2$ there will be 512 (9 feasible states per PEV). Moreover, all the feasible transitions among those combinations between time steps have to be considered, and their optimal paths and associated costs have to be stored to proceed with the FDP algorithm. Besides the increase on memory requirements and computational time, the issues associated to centralized algorithms, mentioned on the conclusions of Chap. 2, have to be considered as well. Thus, to provide a link between multiple PEVs, and optimally manage their charging schedules, a Game Theory approach is proposed in details, on Sect. 3.4. This approach is considered to decentralize the optimization procedures, reduce the memory and computational time requirements, and avoid the flexibility issues of fully centralized optimization approaches.

3.3.3 A Descriptive Example of the State Space

Before passing to the game theory description linking local forward dynamic programming algorithms in a decentralized optimization scheme, it is important

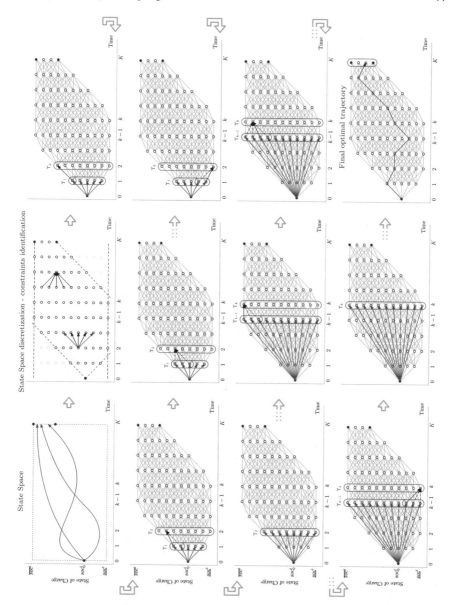

Fig. 3.3 Adaptation of the problem of one PEV charging schedule to the Forward Dynamic Programming Algorithm

Fig. 3.4 Adaptation of the
problem of one PEV
charging schedule to the
Forward Dynamic
Programming Algorithm

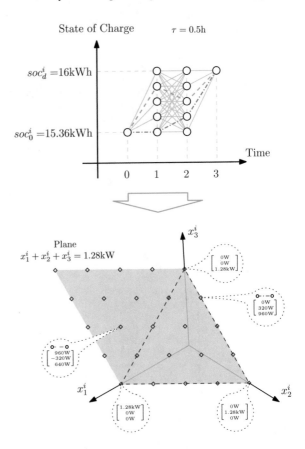

to consider the example of Fig. 3.4. This example is useful to establish a bridge
between the approach proposed in this chapter, and more flexible approaches that
will be described on Chaps. 4, and 5. Besides, it is useful to associate the FDP algorithm and the discretization of the state space, with the concept of strategy applied
in the game theory approach followed on Sect. 3.4.

In this simple example, a PEV has to consume 640Wh to reach a desired state of
charge of $soc_d^i = 16$ kWh. It has $K^i = 3$ time steps of half hour (a horizon of one hour
and a half) to consume this energy, and during this period of time, it cannot discharge
the battery below the initial state of charge ($soc_0^i = 15.36$ kWh). It is assumed that
the state space is discretized with energy steps of 160Wh, or equivalently power
steps of 320 W. Moreover, the charger is considered to be bidirectional, with 3.2 kW
nominal power. Considering this, there are 5 feasible states that could be visited at
$k = 1$, other 5 feasible states at $k = 2$, and at $k = 3$ the desired state must be reached.
Taking this into account, there is a total of 25 feasible trajectories that the state of
charge may follow. Now, let us recall that all of these trajectories are the possible
sums of the energies consumed at each of the three time steps. In other words, let us
recall that,

$$soc_0^i + \tau(x_1^i + x_2^i + x_3^i) = soc_d^i,$$

$$x_1^i + x_2^i + x_3^i = \frac{soc_d^i - soc_0^i}{\tau} = \frac{640\text{Wh}}{0.5\text{h}},$$

$$x_1^i + x_2^i + x_3^i = 1.28\,\text{kW}.$$

Thus, each one of the 25 feasible trajectories of the state variable corresponds to a point in the plane $x_1^i + x_2^i + x_3^i = 1.28\,\text{kW}$ as it is shown of Fig. 3.4. For instance, five examples are highlighted:

- The trajectories where all the required energy is consumed entirely in only one of the three steps:

$$\begin{bmatrix} x_1^i \\ x_2^i \\ x_3^i \end{bmatrix} = \begin{bmatrix} 1.28\,\text{kW} \\ 0\,\text{W} \\ 0\,\text{W} \end{bmatrix}, \quad \begin{bmatrix} x_1^i \\ x_2^i \\ x_3^i \end{bmatrix} = \begin{bmatrix} 0\,\text{W} \\ 1.28\,\text{kW} \\ 0\,\text{W} \end{bmatrix}, \quad \begin{bmatrix} x_1^i \\ x_2^i \\ x_3^i \end{bmatrix} = \begin{bmatrix} 0\,\text{W} \\ 0\,\text{W} \\ 1.28\,\text{kW} \end{bmatrix}.$$

- A trajectory marked with an orange dashed line, where the battery is charged up to 15.84 kWh during $k = 1$. Then it is discharged down to 15.68 kWh during $k = 2$. and finally it is again charged up to the desired state of charge during $k = 3$. This trajectory corresponds to:

$$\begin{bmatrix} x_1^i \\ x_2^i \\ x_3^i \end{bmatrix} = \begin{bmatrix} 960\,\text{W} \\ -320\,\text{W} \\ 640\,\text{W} \end{bmatrix}.$$

- A trajectory marked with a magenta dashed-dotted line, where the state of charge is not changed during $k = 1$. Then it is charged up to 15.52 kWh during $k = 2$. and finally it is again charged up to the desired state of charge during $k = 3$. This trajectory corresponds to:

$$\begin{bmatrix} x_1^i \\ x_2^i \\ x_3^i \end{bmatrix} = \begin{bmatrix} 0\,\text{W} \\ 320\,\text{W} \\ 960\,\text{W} \end{bmatrix}.$$

For the purpose of the game theory approach on the next section, these 25 feasible trajectories will be interpreted as the set of strategies that a player i (representing a PEV with the conditions of this example) would be able to play on the game. For this PEV, It can be observed that the whole set of strategies lies in the plane $x_1^i + x_2^i + x_3^i = 1.28\,\text{kW}$, which is immersed on a 3-dimensional space. In a general case, the total amount of feasible trajectories or strategies will be dependent on the constraints of the PEV and the length of the charging period. Thus, in the general case, the set of strategies will lie on a K^i-dimensional hyper-plane ($\sum_{k=1}^{K^i} x_k^i = (soc_d^i - soc_0^i)/\tau$). These concepts are very important for the purpose of these methods and will be further explored and exploited by the approaches on Chaps. 4 and 5.

3.4 Management of Multiple PEVs: Forward Dynamic Programming Algorithm and Potential Game Approach

To provide a decentralized approach to manage the charging operation of multiple PEVs, a N-person non-cooperative game approach is formulated. In this approach, a number J of players is considered, where each PEV is considered as a player. Each player is labelled with $i = \{1, 2, \cdots, i, \cdots, J\}$, and the set of possible strategies for player i is defined as the set of all possible charging profiles $\mathbf{x}^i = [x_1^i, x_2^i, \cdots, x_k^i, \cdots, x_{K_i}^i]$. On the other hand, for player i, the strategies chosen by the $J - 1$ players left are grouped in the following expression,

$$\mathbf{x}^{-i} = [\mathbf{x}^1, \mathbf{x}^2, \cdots, \mathbf{x}^{i-1}, \mathbf{x}^{i+1}, \cdots, \mathbf{x}^J]$$

The payoff function for player i, depends on the strategy it chooses, and the strategies chosen by other players. This payoff functions is defined as the negative of the squared euclidean distance between the total load at each step of time and the average load,

$$G_i(\mathbf{x}^i, \mathbf{x}^{-i}) = -\sum_{k=1}^{K} \left(\left(x_k^i + \sum_{j=1, j \neq i}^{N} x_k^j + l_k \right) - l_{avg} \right)^2, \qquad (3.6)$$

where, l_k is the forecast of the grid's load (without PEVs) at the transformer, and l_{avg} is the average load during the whole charging period including grid's base load and PEVs' load. The average load l_{avg} is given by,

$$l_{avg} = \frac{\sum_{i=1}^{J} \left(soc_d^i - soc_0^i \right) / \tau + \sum_{k=1}^{K} l_k}{K}.$$

To simplify the payoff expression, let us regroup some terms in,

$$l_k^{-i} = -\sum_{j=1, j \neq i}^{J} x_k^j - l_k + l_{avg}. \qquad (3.7)$$

Then, the payoff for player i is given by,

$$G_i(\mathbf{x}^i, \mathbf{x}^{-i}) = -\sum_{k=1}^{K^i} \left(x_k^i - l_k^{-i} \right)^2. \qquad (3.8)$$

It is important to notice that these utility functions are strictly concave. Now, assuming that the other players have chosen their best strategies $\mathbf{x}^{-i*}$, the *best reply* (BR) strategy $\mathbf{x}^{i*}$ for player i is given by,

$$\mathbf{x}^{i*} \in \arg \max_{\mathbf{x}^i \in \Theta^i} G_i(\mathbf{x}^i, \mathbf{x}^{-i}). \tag{3.9}$$

The set Θ^i is defined by the constraints detailed in Sect. 3.2. Given the fact that these constraints are linear, the set Θ^i is convex, closed and bounded. To clarify the concept behind the set of strategies Θ^i for player i, an illustrative example was introduced on Sect. 3.3.3, using Fig. 3.4.

Given that function (3.6) represents the payoffs for all J players, and it can be employed to obtain their best-replies (3.9), this game is a *best-response potential game* [Voo00, DLP08], with potential function $G_i(\mathbf{x}^i, \mathbf{x}^{-i})$ given by(3.6). In this type of games, best-reply algorithms like the one that will be explained, can be applied to find an equilibrium.

The *Nash equilibrium* is defined as a state (or vector of strategies $\mathbf{x}^* = [\mathbf{x}^{i*}, \mathbf{x}^{-i*}]$) in which no player can improve its payoff by unilaterally deviating from its equilibrium strategy $\mathbf{x}^{i*}$. In other words, in the Nash equilibrium state, if one player deviates or changes its strategy from the equilibrium strategy (while other players maintain their equilibrium strategies), then its payoff can only be reduced [Web07, HS88, NS12].

Given the strict concavity of utility functions for each player and the fact that sets Θ^i are convex, closed and bounded, the game is a strictly concave N-person game where the Nash equilibrium exists and it is unique [Ros65, NS12, Wu+11].

The approach followed to find the Nash equilibrium is an algorithm similar to the one proposed by [NS12]. Each player receives an updated total load profile λ, including all other players' strategies and grid load, from a centralized agent or aggregator. The updated total load profile can be expressed as,

$$\lambda_k = l_k + \sum_{i=1}^{J} x_k^i. \tag{3.10}$$

Based on the total load, the player solves the local optimization problem (with the Forward DP algorithm) to find its BR strategy $\mathbf{x}^{i*}$, and if it is different from its previous BR strategy $\mathbf{x}^i$, the player sends its updated BR strategy to the aggregator. The algorithm[1] is summarized in Fig. 3.5. As it is proved by authors of [NS12], this best response algorithm converges to a Nash equilibrium if players choose their best responses (maximize their payoff) in a sequential and asynchronous fashion.

The role of the aggregator in this algorithm is the compilation of information from players and the transmition of updated information to each player sequentially. Players interact with other player only through the aggregator. Additionally, each local FDP algorithm bases its optimal charging scheduling procedure on the computation of a short-term load forecast l of the transformer's load. This kind of forecast is possible with appropriate stored historical measurements [PS13, Bas+15].

---

[1]To shorten the notation on the algorithm description of Fig. 3.5, states are represented in short-form by s_k^i instead of soc_k^i.

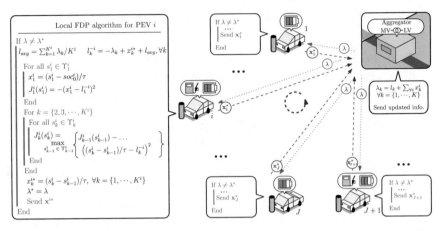

Fig. 3.5 Best-reply structure for the proposed PEVs N-person game. Decentralization of scheduling procedures

The cost function of transitions between states, for the DP algorithm, is defined based on the payoff function of the N-person game as follows,

$$g_k^i = -(x_k^i - l_k^{-i})^2 = -\left(\frac{soc_k^i - soc_{k-1}^i}{\tau} - l_k^{-i}\right)^2, \qquad (3.11)$$

If players update their BR strategies in an asynchronous fashion, their payoffs will either increase or remain the same. Since the distance to the average load is bounded below (in the best case it could be zero), then the algorithm will converge to the desired point or the Nash equilibrium [NS12].

3.5   Illustrative Example with a Reduced Number of PEVs

To evaluate the performance of the charging management approach of this chapter, a test case is proposed with 10 PEVs having the following characteristics. All of them have battery capacities of 20 kWh, and their state of charge profiles are constrained to be between 30 and 80% of the capacity (6–16 kWh). States of charge are discretized with energy steps of 80 Wh. Thus, the whole range is divided in 126 steps from 6 to 16 kWh, both limits included.

Chargers are considered to be bidirectional with charging rates of going from -3.2 to $+3.2$ kW ($\overline{p}^i = 3.2$ kW). With energy steps of 80 kWh, and time steps $\tau = 0.25$ h, the resulting step size of input/output power is 320 W. Thus, the range of power is divided in 21 steps including 0 W ad the upper and lower limits.

The period of charge is chosen between 18h in the evening and 06h in the morning. During this period of time, the forecasted grid load without PEVs is shown

Fig. 3.6 Total load profile without PEVs. © [2017] IEEE. Reprinted, with permission, from [OHB15]

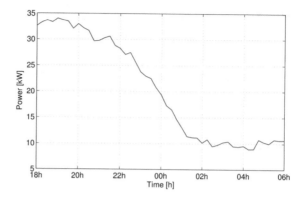

on Fig. 3.6. The initial states of charge for each PEV are chosen randomly with a uniform probability distribution between 30 and 40%. The resulting initial states of charge are 37.6, 36.8, 33.6, 40.0, 31.2, 38.8, 34.4, 36.4, 37.6, and 37.2%.

Given these initial states of charge, the initial strategies for each player are defined with the policy of uniformly charging during the whole period of charge (uniform charging rates for all PEVs during the charging period). The uniform rate of consumption is defined by,

$$x_k^i = \frac{soc_d^i - soc_0^i}{\tau K^i},$$

depending on the PEV, where K) is the total number of time steps in the charging period of each PEV. In this test case, all PEVs have the same $K^i = 49$.

Given these initial strategies, the initial load curve including grid and PEVs is shown on Fig. 3.7. As it can be observed, this initial profiles has the same shape of the grid's load forecast, with an offset that corresponds to the PEVs load. The subsequent updates of each PEV are also shown on this figure. It is important to notice that once the last PEV updates its strategy, the final load curve does not change much since the Nash equilibrium is almost reached. Also, it is important to notice how the total load curve becomes flatter with the progress of the game. This occurs because the utility of each player is higher if the total load at each step of the changing period becomes closer to the average load. In other words, given the potential game nature of the approach, each time each PEV updates its best response, all other PEVs improve their payoffs as well.

Equilibrium strategies (i.e. the final optimal power consumption schedules) are shown on Fig. 3.8. On the other hand, state of charge profiles are shown on Fig. 3.9. It is important to notice how each player respects the constraints of minimum state of charge (30%), final desired state of charge (80%), and power limits (±3.2 kW).

Figure 3.10 shows a comparison between the grid load, the optimal total load (grid and PEVs' load), the average load, and the aggregated load of PEVs. It is important to notice that the consumption of PEVs is mostly moved to off-peak hours (valley filling). During peak demand hours, PEVs tend to discharge their batteries (if energy

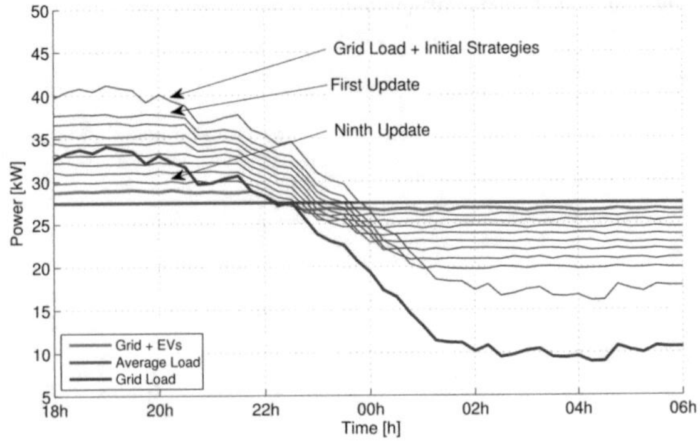

Fig. 3.7 Evolution of the game with 10 PEVs. © [2017] IEEE. Reprinted, with permission, from [OHB15]

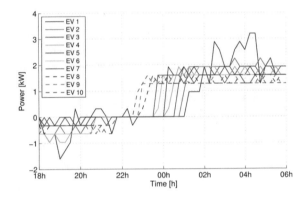

Fig. 3.8 Optimal power consumption profiles, final strategies in the Nash equilibrium. © [2017] IEEE. Reprinted, with permission, from [OHB15]

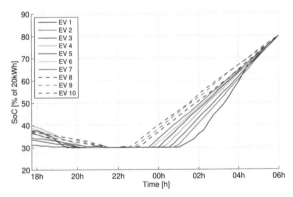

Fig. 3.9 Optimal charging schedules, states of charges for each player. © [2017] IEEE. Reprinted, with permission, from [OHB15]

Fig. 3.10 Optimal power consumption profiles, final strategies in the Nash equilibrium. © [2017] IEEE. Reprinted, with permission, from [OHB15]

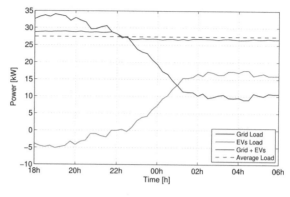

Fig. 3.11 Optimal charging schedules, states of charges for each player. © [2017] IEEE. Reprinted, with permission, from [OHB15]

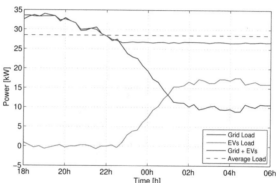

is available) to provide peak reduction. If the initial states of charge are close to the lower limit (30% in this example), batteries can not be discharged and PEVs cannot provide peak compensation. This case can be observed in Fig. 3.11. However, PEVs are still charged mostly during off-peak hours.

3.6   Test Case with the IEEE European Low Voltage Test Feeder

In this section, an additional study case is considered to compare the proposed decentralized scheduling approach of this chapter, in contrast with the centralized scheduling methodology presented on Chap. 2. The test in performed under the same realistic conditions proposed on Sect. 2.5. As it was mentioned on that chapter, in this scenario the IEEE European Low Voltage Test Feeder is considered [Iee, Ope]. Besides, several PEV chargers are placed on different buses of the grid, with or without residential load, as it is shown on Fig. 3.12. In total there are 33 chargers with nominal powers of 3.3 kW and 7.5 kW distributed through out the grid.

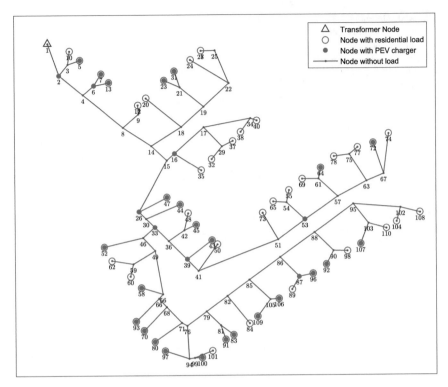

Fig. 3.12 Random allocation of PEV chargers on the buses of the simplified version of the IEEE European Low Voltage Test Feeder

Let us recall the features of the study case. PEVs arrive in a random way, following a *Poisson* model with variable rates of arrival according to the hour of the day, and variable connection times. The highest rate of arrivals is 10 PEVs/h at 17 h and it decays up to 0.5 PEVs/h at 16 h on the next day. Vehicle battery capacities can be 8.8 kWh with a probability of 30%, and 20 kWh with a probability of 70%. Furthermore, chargers can have nominal powers of 3.3 kW with a probability of 80%, and 7.4 kW with a probability of 20%. To reduce the impact on the battery life span, states of charge are limited to 25–85% for 8.8 kWh capacities (2.2–7.5 kWh), and 30–80% for the 20 kWh capacities (6–16 kWh). A summary of these assumtion can be found on Table 3.1.

The number of arrivals and departures each half hour, can be checked on Fig. 3.13, in contrast with:

- The forecast of the grid's load profile.
- The total number of connected PEVs.
- The number of PEVs connected to 3.3 and 7.5 kW chargers
- The number of PEVs with capacities of 8.8 and 20 kWh.

Table 3.1 Descriptive summary and assumptions of the considered study cases

Item	Description
Chargers	3.3 kW with probability of 80%
	7.5 kW with probability of 20%
Batteries	8.8 kWh with probability of 30%
	20 kWh with probability of 70%
Constraints on batteries	Between 25% and 85% for 8.8 kWh
	Between 30% and 80% for 20 kWh
Highest rate of arrivals	10 PEVs/h at 17h
Lowest rate of arrivals	0.5 PEVs/h at 16 h next day
Peak of connected PEVs	30 PEVs at 01 h in the morning
Distribution system info.	IEEE European low voltage test feeder
Evaluated scenarios	Comparison with centralized approach of Chap. 2 with double tariff.
Time period	24 h (half hour steps)

It can be observed that most of the PEVs arrive during hours of peak consumption in the afternoon. Besides, there is a large amount of PEVs connected during low demand hours. As it can be expected from the previously mentioned probabilities, the most common nominal power of chargers is 3.3 kW, while most of the PEVs have batteries of 20 kWh.

The purpose of this test case is to consider a similar scheduling scenario to that of Sect. 2.5. However, with the decentralized optimization approach of this chapter, the objective is to avoid the use of a distribution grid model. Instead of collecting multiple local load profile forecasts to feed a grid model in a centralized scheduling routine, in this case only the forecast of the transformer load is employed. Even if there is an information exchange between a central aggregator and PEVs, the idea behind the FDP/Potential Game approach is to distribute optimization procedures on each PEV charger controller. Furthermore, each PEV receives the aggregated version of the total load, including both the forecast of the grid's base load and the PEVs' load. In other words, individual PEVs do not have specific knowledge of the schedules of other PEVs, to avoid privacy issues.

Based on these profiles of PEVs arriving and departing, let us recall the results obtained by the centralized approach on Sect. 2.5.5. In that section, two different tariffs were considered, one for day hours and a smaller one for night hours. The lower tariff was applied between 22 and 06 h in the morning the next day. Initially, voltage constraints were fixed uniformly through out the evaluated day, between $v_{min} = 0.95 V_{nom}$, and $v_{max} = 1.05 V_{nom}$. It was evidenced that batteries are discharged as much as possible before the beginning of low tariff hours. Then, PEVs consume as much as possible, such that the total cost of charging is as lowest as possible. As a result of this greedy effect, stronger peaks of consumption are likely to occur in the borders between low and high tariff hours, as it was observed on Fig. 2.28.

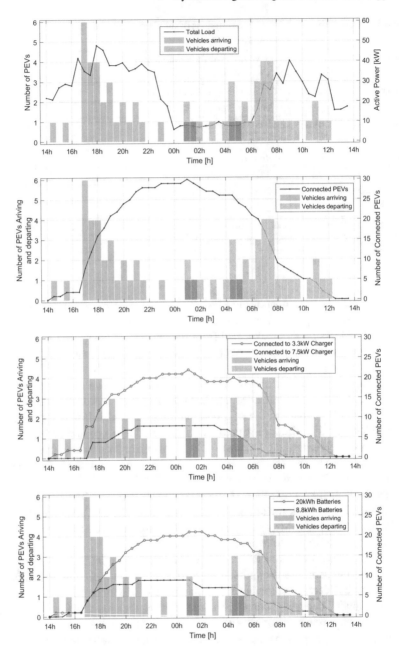

Fig. 3.13 Profiles of arrivals and departures of PEVs during the test day, in contrast with: a forecast of the total load profile of the grid, the resulting total amount of connected PEVs, the total amount of connected PEVs to 3.3 and 7.5 kW chargers, and the total amount of PEVs with battery capacities of 20 and 8.8 kWh

To mitigate the issues caused by the double tariff in the centralized approach, voltage limits were modified such that PEVs are still economically motivated to consume during low tariff hours, without creating new strong peaks of consumption. Voltages were constrained between $v_{min} = 1.03V_{nom}$, and $v_{max} = 1.05V_{nom}$, during low tariff hours, and between $v_{min} = 1.01V_{nom}$, and $v_{max} = 1.05V_{nom}$ elsewhere. The scheduling results can be observed again on Fig. 3.14. These constraints put boundaries on the greedy effect of the double tariff, and force PEVs to limit the amount of energy allocated during low tariff hours. As it can be observed, PEVs take as much advantage of the low tariff as they can since most of the voltages are on the edge of the lower constraint during this period. On the other hand, constraints during high tariff hours have the effect of forcing PEVs to avoid load allocation during peaks of consumption, and even reallocate those peaks using the batteries. Consequently, it can be observed that the original peak at 18h, is reduced and it becomes similar to the peaks of consumption after 06h.

On the other hand, Fig. 3.15 shows the results obtained by the decentralized optimization approach of this chapter, under similar conditions. Several details can be noticed by comparing with the approach of Chap. 2. On a first place, given that the global objective (i.e. the potential function) is to reduce the distance between the total load profile and the average load of the day, vehicles automatically use their initial stored energy soc_0^i to reduce the base load peaks between 16 and 20h. This compensation is more pronounced after 17h where the peak of arrivals of PEVs occurs. During the peak of consumption in the morning at 09h, most PEVs are close to leave and some of them are not able to inject energy to the grid. However, three 20 kWh PEVs are programmed to leave around 12h, so they have enough time to inject energy and compensate the peak at 09h, and charge again their batteries up to the desired states of charge. Moreover, it is observed that the objective function naturally motivates PEVs to consume during low demand hours.

In terms of the voltage profiles, it can be observed that the flattening compensation that PEVs have over the total load profile (base load forecast and PEVs' load) at the transformer node, has a proportional flattening effect on the voltage profiles of each nodes of the grid. This occurs because the impedance of lines (i.e. the electrical distance between nodes) on electrical distribution systems is relatively small [BV00, Gue+05, Ova+15]. As a consequence, the effect of load variations over time on a node of the grid is likely to affect the voltage profiles of all the nodes.

Finally, on the state of charge profiles of Fig. 3.15 it is possible to observe that upper and lower boundaries are respected though out the charging period of each PEV, and final desired states of charge are reached at the end of the charging periods as well. However, under both centralized and decentralized optimization approaches, it is possible to observe that resources are not fairly allocated among PEVs. Some vehicles are likely to compensate base load peaks and valleys more than others, and some of them are likely to get fully charged earlier than others. In other words, both methodologies behave like *first come—first served* policies.

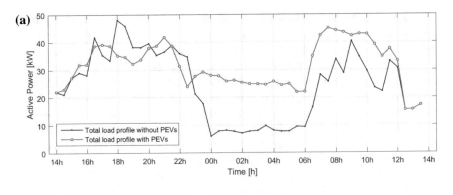

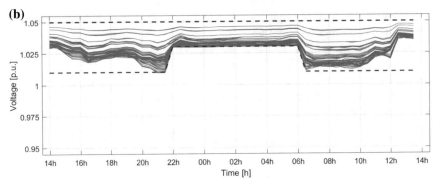

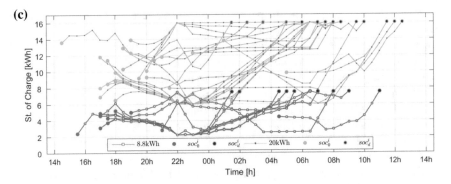

Fig. 3.14 Results for the centralized approach of Chap. 2, with double tariff and bidirectional chargers. **a** Total load profiles with and without PEV load; **b** Resulting voltage profiles with PEVs after optimal scheduling; **c** Resulting optimally scheduled state of charger profiles

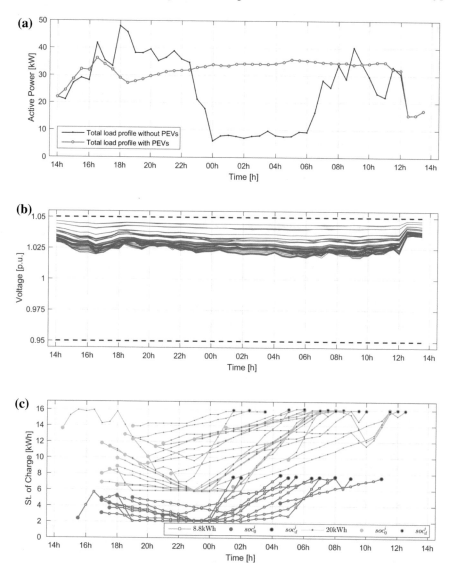

Fig. 3.15 Results for the decentralized FDP/POtential game approach of this chapter. **a** Total load profiles with and without PEV load; **b** Resulting voltage profiles with PEVs after optimal scheduling; c Resulting optimally scheduled state of charger profiles

3.7   Matlab Scripts

The following Matlab script allows you to test the forward dynamic programming algorithm employed in this chapter for each local scheduler in the game theory approach. We encourage the reader to modify it and adapt it to its needs. It implements the routine described using Fig. 3.3.

```
function [pow_traj_ast,soc_traj_ast]= Book_DP_Forward(soc_c,soc_d,soc_0,...
                                  p_max,dt,Ki,Lhat_,x0)

% ------ Desciption inputs
% soc_c : battery available capacity [kWh]
% soc_d : final desired state of charge [kWh]
% soc_0 : initial state of charge [kWh]
% p_max : max charging/discharging rate [kW]
% dt : time step
% Ki : total time steps
% Lhat_ : Load forecast
% x0 : previously defined charging rate trajectory

% ------ Description outputs
% pow_traj_ast : optimal power trajectory
% soc_traj_ast : optimal soc trajectory


L_others = Lhat_ - x0;
kwh_discr = 10; % kwh_discr+1  is the maximum amount of states at a given
                % time

kwh_max = p_max*dt;
kwh_step = kwh_max/kwh_discr;

soc_ini = ceil(soc_0/kwh_step)*kwh_step;
soc_max = ceil(soc_c/kwh_step)*kwh_step;
soc_des = ceil(soc_d/kwh_step)*kwh_step;
soc_min = 0;

costs_optimal_states = [];

aux1 = soc_des - Ki*kwh_max;
aux2 = soc_des + Ki*kwh_max;

for tn2 = 1:Ki % loop through time steps

    tn1 = tn2-1;

    % Smallest soc for tn2
    soc_tn2_lim = max([soc_min,soc_ini-tn2*kwh_max,aux1+tn2*kwh_max]);

    % Number of possible states at tn2
    s2 = 1+(min([soc_max, soc_ini+tn2*kwh_max, aux2-tn2*kwh_max])...
        -soc_tn2_lim)/kwh_step;

    % Smallest soc for tn1
    soc_tn1_lim = max([soc_min,soc_ini-tn1*kwh_max,aux1+tn1*kwh_max]);

%      % Number of possible states at tn1
```

```
%      s1 = 1+(min([soc_max,soc_ini+tn1*kwh_max])-soc_tn1_lim)/kwh_step;

%%%%%%%%%%%%%%%%%%%%%%%%%%%%%%%%%%%%%%%%%%%%%%%%%%%%%%%%%%%%%%%%%%%%%%%%%%%%

costs_optimal_states_pre = costs_optimal_states;
costs_optimal_states = zeros(s2,3+tn1);

for i = 1:s2 % loop over states in tn2
    % Recovers Wh value for state i in tn2
    soc_tn2 = soc_tn2_lim+kwh_step*(i-1);

    % Smallest root state in tn1 for state i in tn2
    soc_r_tn1_lim = max([soc_min,soc_ini-tn1*kwh_max,...
        aux1+tn1*kwh_max,(soc_tn2-tn2*kwh_max)+tn1*kwh_max]);

    % amount of root states in tn1 for state i in tn2
    s1_r = 1+(min([soc_max,soc_ini+tn1*kwh_max,...
        aux2-tn1*kwh_max,(soc_tn2+tn2*kwh_max)-tn1*kwh_max])...
        -soc_r_tn1_lim)/kwh_step;

    % Loop over all possible root states for the state in tn2 under
    % evaluation. At the same time, the best option is evaluated and
    % and its index is kept

    optimal_cost = Inf;
    optimal_root_state = 0;

    for j = 1: s1_r % s1

        % Recovers Wh value for root state j in tn1
        soc_r_tn1 = soc_r_tn1_lim+kwh_step*(j-1);

        % Recovers general index Aj for root state j in tn1 (Aj is not
        % equal to j)
        % Index j is only valid for the root states of the state of tn2
        % under evaluation

        Aj = (soc_r_tn1-soc_tn1_lim)/kwh_step+1;

        if tn1 == 0
            % charge or discharge rate
            p=(soc_tn2-soc_r_tn1)/dt;
            cost = (L_others(tn2)+p)^2;
            optimal_cost = cost;
            optimal_root_state = Aj;
        else
            % Checking for existence of the root state among
            % previously saved optimal states in tn1

            index = find(costs_optimal_states_pre(:,1)==Aj);
            if isempty(index)
                % root state is ignored
            else
                % charge or discharge rate
                p = (soc_tn2-soc_r_tn1)/dt;
                cost = costs_optimal_states_pre(index,2) + ...
                    (L_others(tn2)+p)^2;
                if cost < optimal_cost
                    optimal_cost = cost;
```

```
                              optimal_root_state = Aj;
                    end
                end
            end
        end

        % Register valid states in tn2 and their associated optimal root
        % state and cost
        if optimal_cost < Inf
            if tn1 == 0
                index0 = find(costs_optimal_states(:,1)==0);
                costs_optimal_states(index0(1),1:3) = [i, ...
                                   optimal_cost,optimal_root_state];
            else
                index0 = find(costs_optimal_states(:,1)==0);
                index1 = costs_optimal_states_pre(:,1)==optimal_root_state;
                EstadosPreviosOptimos = ...
                    costs_optimal_states_pre(index1,3:end);
                costs_optimal_states(index0(1),1:3+tn1) = ...
                    [i,optimal_cost,...
                    EstadosPreviosOptimos,optimal_root_state];
            end
        else
            costs_optimal_states(end,:) = [];
        end
    end
    costs_optimal_states(costs_optimal_states(:,1)==0,:)=[];

end

optimal_policy=costs_optimal_states(3:end);

soc_traj_ast=0*optimal_policy;
for k = 1:Ki
    t = k-1;
    % smallest state of charge in t
    soc_lim = max([soc_min,soc_ini-t*kwh_max,aux1+t*kwh_max]);
    soc_traj_ast(k) = soc_lim+kwh_step*(optimal_policy(k)-1);
end

% output assignments
soc_traj_ast = [soc_traj_ast soc_des];
pow_traj_ast = (soc_traj_ast-[soc_ini soc_traj_ast(1:end-1)])/dt;
pow_traj_ast = pow_traj_ast(2:Ki+1);

end
```

3.8 Conclusion

This chapter proposes a Dynamic programming—Potential game approach for
obtaining the optimal charging schedules of multiple PEVs in a decentralized optimization scheme. Additionally, the proposed strategy employs the energy storage
capacity of the PEVs to provide a service of load flattening compensation. The ideas
behind the concept of strategy in the potential game are analyzed and associated to

the conditions of each PEV (initial state of charge, desired state of charge, expected departure time, and nominal power of the charger), and the procedure followed by each local forward dynamic programming algorithm.

The proposed approach is evaluated under multiple study cases to test its capabilities. Besides, it is compared with the previously proposed centralized approach of Chap. 2, under realistic conditions: IEEE European Low voltage test feeder, stochastic PEV arrivals and departures, different types of charges and battery capacities. These results are useful to conclude that instead of centralizing information and optimization procedures, their decentralization allows to tackle the optimal scheduling of PEV load in a much more flexible way, reducing the information requirements.

It has been observed that this kind of approaches where DP algorithms are employed rely on the discretization of the state space. Consequently if more that one state or control variable has to be considered, as it is for instance with three-phase chargers, then performance is severely affected due to combinatorial issues. However, besides the conclusion on the decentralization of optimization procedures and its flexibility, it is important to highlight the results analyzed on Fig. 3.4. Here the concept of strategy was illustratively associated to the plane or the hyper-plane where they belong. This illustration is very important for the more flexible and elaborated approaches of the following chapters.

Chapter 4
Evolutionary Game Theory Approach Part I: Mixed Strategist Dynamics

In this chapter, the first part of an evolutionary game theory approach for decentralized PEV load scheduling is presented. This approach is based on a multi-population scenario modeled using an evolutionary game dynamics called Mixed Strategist Dynamics (MSD). First, a theoretical description of the MSD is given with some numerical examples motivating this application. Then, the multi-population scenario for the PEV load scheduling problem is presented. In this scenario, shares of populations distributed over multiple territories evolve depending on the payoffs that territories provide. In this context, individual populations represent PEV loads, and these integrate a total population representing the transformer load. Territories are pure strategies for populations in the evolutionary game. Then, territories represent time slots where energy consumption (i.e., population shares) can allocated. The distribution of each single population evolves following the MSD according to a payoff provided for using mixes of pure strategies (mixed strategies). These mixed strategies are exploited in the application to represent charger's constraints. In this approach, PEV owners can be economically or socially motivated by the utility grid manager to participate in the scheduling process. The concept of *fairness* in resource allocation of connected PEVs is introduced. Performance is evaluated using SOREA utility grid company real data. The chapter ends with Matlab scripts for obtaining mixed strategies sets, for the local MSD routines, and a final script representing a simulation scenario used in the chapter. This final script is provided to allow the reader to test the proposed MSD approach and some results of the chapter.

4.1 Introduction

Evolutionary game dynamics is a research field that represents the intersection between population dynamics methods and game theory [HS88, Pel09]. Some well-known dynamics are the Projection dynamics, the Logit dynamics, the Best reply (BR) dynamics, the Replicator dynamics (RD), and in particular the Mixed Strategist dynamics (MSD) employed in this chapter [HS88]. RD has been applied as a tool to solve NP-hard combinatorial optimization problems [Pel09, Men01]. Recently, in the

© Springer International Publishing AG 2018
A. Ovalle et al., *Grid Optimal Integration of Electric Vehicles: Examples with Matlab Implementation*, Studies in Systems, Decision and Control 137, https://doi.org/10.1007/978-3-319-73177-3_4

energy and power systems domain, authors of [PQ11, PQP14] introduced the RD as a tool for solving the optimal decentralized power dispatch of distributed generators in a microgrid topology, under constraints on the communication infrastructure. While RD considers pure strategists, MSD describes the evolution of a population following sets of predefined mixed strategies. This approach takes advantage of this feature of the MSD to model local constraints involved in the problem of charging multiple PEVs in a distribution grid, and include other owner-side desirable constraints.

This chapter describes an application of the MSD proposed for the decentralized PEV load scheduling problem [Ova+16a, Ova+16b]. The proposed formulation is such that the maximum entropy principle (MEP) is applied to achieve load distributions as flat as possible given the constraints imposed by owners and chargers. Entropy measurements of the total load distribution and local PEV load distributions are considered as objective functions, and a trade-off among them is defined by PEV owners. Thus, the role of PEV owners in the scheduling process becomes relevant, and the utility grid manager is able to influence their participation economically or socially.

While entropy maximization for the local load distributions over time contributes to preserve the batteries' states of health, entropy maximization for the total load distribution over time reduces the effects of load peaks on the transformer, and limits strong variations on the total load. The problem is formulated such that final states of charge are assured depending on time constraints defined by owners. An aggregator is in charge of receiving local load distributions from PEVs, add them to the forecast base load, and redistribute the updated information to PEVs. Further details and evaluation scenarios are detailed in the chapter.

It should be noticed that the considered load distributions are points in a simplex manifold similar to the standard simplex manifold where discrete probability distributions lie in. In the statistical manifold, the Shannon's entropy of a distribution is maximized when it is uniform [KK89]. In this proposed approach, this property is proved for the considered entropy functions. Following this analogy with probability distributions, the benefits that populations find in MSs are proposed to be defined by an entropy measure of the total load distribution (including base load and PEVs) over a forecast time horizon, and the entropy measure of the local population distribution over time slots allowed by the PEV owner. Thus, the evolution of populations given by the MSD is such that these entropies are maximized in the trade-off defined by PEV owners, given incentives from the utility grid manager. Given the convex nature of both entropy measures (i.e. the existence of a global attractor), the MSD convergence is guaranteed [HS88].

This approach, originally discussed in [Ova+16a], is explained such that it can be adapted to other kind of decentralized resource allocation scenarios where several control parameters and local constraints arise [JTG13, XL15, Liu+15, GM15, Ova+15]. A brief description of the population analogies employed in this multi-population scenario application are the following. A single population represents the total energy required by a PEV to fully charge its battery. Thus, portions of each population must choose among mixed strategies (MSs). MSs are convex combinations of pure strategies, and pure strategies represent time slots where energy consumption can be allocated (or making the analogy, spots where portions of populations can be allocated). Then, portions of several populations may choose one time slot to *settle* in. MSs are considered for several reasons, for instance, to limit the size of portions of populations able to *settle* in a given time slot. These analogies are further detailed on Sect. 4.3.

Finally, this chapter provides Matlab scripts for the local MSD routines employed by each PEV. Additionally, a script representing a realistic scenario with multiple PEVs arriving and leaving during multiple days is provided as well. The reader is able to use and adapt these scripts to different scenarios, or even increase their complexity.

4.2 Introduction to the Mixed Strategist Dynamics

Let us consider a normalized population represented by a state vector $\mathbf{x} \in \mathbb{R}^K$. Its k-th component x_k represents the portion of that population that may choose a pure strategy k. Thus, the population is divided in K portions choosing one of K pure strategies. Consequently, portions must be $x_k \geq 0$, $\forall k$, and the sum of portions of population must be equal to the unity, $\sum_{k=1}^{K} x_k = 1$.

4.2.1 Preliminaries

For visualization, if $\mathbf{x}$ is interpreted as the vector of coefficients in a linear combination of the column vectors of the canonical base $\mathbf{I}$ of $\mathbb{R}^K$, then the population state vector $\mathbf{x}$ represents by itself a point in the standard simplex Δ^K, formally defined by,

$$\Delta^K = \left\{ \mathbf{x} \in \mathbb{R}^K : x_k \geq 0, \sum_{k=1}^{K} x_k = 1 \right\}.$$

For instance, the standard simplex in $\mathbb{R}^3$ is the triangle whose vertices are the canonical vectors, and $\mathbf{x}$ lies in that triangle for any value of its components, as in Fig. 4.1. In fact, this linear combination is a convex combination given the constraints on the coefficients.

Let us assume that the original population can be divided in a different amount of portions M, where each portion is pre-programmed to use one of M convex combinations of the original K pure strategies. These M convex combinations are called mixed strategies. Now, the same population is represented by a different state vector $\mathbf{y} \in \mathbb{R}^M$, whose components also must be $y_m \geq 0$, $\forall m$, and $\sum_{m=1}^{M} y_m = 1$.

Given that the column vectors of the canonical base $\mathbf{I}$ represent the K pure strategies, let us represent the set of M MSs as the column vectors $\mathbf{c}_m$ of a matrix $\mathbf{C}$. Thus, the element $c_{k,m}$ of $\mathbf{c}_m$, is the coefficient in the convex combination corresponding to pure strategy k in the mixed strategy m. Thus, it must satisfy $c_{k,m} \geq 0$, $\forall k$, and,

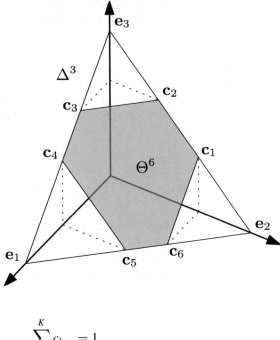

Fig. 4.1 Simplex Δ^3 in $\mathbb{R}^3$. Vectors $\mathbf{c}_m$ represent mixed strategies, and the vertices of a convex set Θ^6, subset of Δ^3

$$\sum_{k=1}^{K} c_{k,m} = 1.$$

Therefore, MSs can be represented also by points $\mathbf{c}_m$ in the interior of the simplex, as it is also shown in Fig. 4.1. Furthermore, according to all the definitions before, the following equivalence holds between state vectors $\mathbf{x}$ and $\mathbf{y}$,

$$\mathbf{x} = \mathbf{C}\mathbf{y}, \tag{4.1}$$

which means that a distribution $\mathbf{y}$ of the population, given a set of MSs, has an equivalent distribution $\mathbf{x}$ in terms of the original pure strategies. On the other hand, let us express the benefits provided by pure strategies by the column vector

$$\mathbf{f}(\mathbf{x}) = [f_1(\mathbf{x}), f_2(\mathbf{x}), \dots, f_k(\mathbf{x}), \dots, f_K(\mathbf{x})]^{\mathsf{T}},$$

where element $f_k(\mathbf{x})$ is the payoff function associated to pure strategy k. Besides, let us express the weighted average payoff of the population given pure strategies as,

$$\bar{f}(\mathbf{x}) = \sum_{k=1}^{K} x_k f_k(\mathbf{x}) = \mathbf{x}^{\mathsf{T}}\mathbf{f}(\mathbf{x}).$$

Given a state vector $\mathbf{y}$, from (4.1) the benefits in terms of pure strategies can be assumed to be $\mathbf{f}(\mathbf{x}) = \mathbf{f}(\mathbf{C}\mathbf{y})$. On the other hand, let us assume that the portion of population y_m using mixed strategy $\mathbf{c}_m$ gets a payoff from pure strategy k proportional to the corresponding coefficient $c_{k,m}$ in the mixed strategy. This means that the total benefit that portion y_m gets from its mixed strategy is $g_m(\mathbf{y}) = \sum_{k=1}^{K} c_{k,m} f_k(\mathbf{Cy})$. Then the payoff vector for MSs is defined by

$$\mathbf{g}(\mathbf{y}) = \mathbf{C}^\mathrm{T}\mathbf{f}(\mathbf{Cy}).\tag{4.2}$$

Given the distribution $\mathbf{y}$, the resulting weighted average payoff of the population is,

$$\bar{g}(\mathbf{y}) = \mathbf{y}^\mathrm{T}\mathbf{g}(\mathbf{y}) = \mathbf{y}^\mathrm{T}\mathbf{C}^\mathrm{T}\mathbf{f}(\mathbf{Cy}) = \mathbf{x}^\mathrm{T}\mathbf{f}(\mathbf{x}) = \bar{f}(\mathbf{x}),$$

which implies that, the weighted average payoff in terms of the population distribution for MSs is equivalent to the weighted average payoff in terms of the corresponding distribution for pure strategies. Based on these definitions, let us introduce the approach of the MSD followed in this chapter.[1]

4.2.2 The MSD Equation

The MSD describes the evolution of M portions of a normalized population following M different MSs defined on K pure strategies, according to the benefit provided by those pure strategies. For portion y_m of the population, the continuous time dynamics can be described by

$$\dot{y}_m = y_m((\mathbf{C}^\mathrm{T}\mathbf{f}(\mathbf{Cy}))_m - \bar{g}(\mathbf{y})).\tag{4.3}$$

From (4.3) it can be noticed that: if portion m is initially zero it will remain zero; portions with payoff greater that the weighted average payoff tend to grow and viceversa; an equilibrium is reached when $g_m(\mathbf{y}) = \bar{g}(\mathbf{y})$. Summing (4.3) over all the portions of the population provides

$$\frac{d}{dt}\sum_{m=1}^{M} y_m = \sum_{m=1}^{M} y_m(\mathbf{C}^\mathrm{T}\mathbf{f}(\mathbf{Cy}))_m - \bar{g}(\mathbf{y})\sum_{m=1}^{M} y_m$$

$$= \bar{g}(\mathbf{y})\left(1 - \sum_{m=1}^{M} y_m\right).$$

Therefore, if at $t = 0$ the sum of portions of population is equal to the total population $\sum_{m=1}^{M} y_m = 1$, then the sum of portions will be equal to the total population for all $t > 0$ since $\sum_{m=1}^{M} \dot{y}_m = 0$. Besides, given that (4.3) sets a lower limit on zero

---

[1]Elements f_k, and g_m are completely different and should not be confused with the nomenclature employed on the dynamic programming algorithm of Chap. 3, for the model of the system and the additive cost function.

to y_m, then $\mathbf{y}$ will always lie within another standard simplex Δ^M, formally defined by

$$\Delta^M = \left\{ \mathbf{y} \in \mathbb{R}^M : y_m \geq 0, \sum_{m=1}^{M} y_m = 1 \right\}.$$

Thus, MSD satisfies (4.1) and the state vector $\mathbf{x}$ will always lie within a subset $\Theta^M \subset \Delta^K$, formally defined by the following convex hull,

$$\Theta^M = \left\{ \mathbf{x} = \mathbf{Cy} \in \mathbb{R}^K : y_m \geq 0, \sum_{m=1}^{M} y_m = 1 \right\}.$$

With proper definition of MSs it is possible to limit the dynamic of the population state vector $\mathbf{x}$ to a reduced subset of the standard simplex Δ^K. This is the first interesting feature of the MSD exploited in this chapter.

To check the effect of MSs, an example with the three pure strategies game, Rock-Paper-Scissors, is presented [HS88]. It is described by payoff functions

$$\mathbf{f}_{rps}(\mathbf{x}) = \begin{bmatrix} 0 & -a & b \\ b & 0 & -a \\ -a & b & 0 \end{bmatrix} \mathbf{x},$$

where $a, b > 0$. This game has a global attractor on $\hat{\mathbf{x}} = [1/3, 1/3, 1/3]^T$ for $a < b$. Three subsets Θ^M defined by different sets of MSs, containing the same global attractor, are presented on Fig. 4.2. The trajectories followed by the MSD, with each set of MSs, are presented. The dark blue trajectory results when the set of MSs is identical to the set of pure strategies, the trajectories can reach any point within the simplex. The orange one is the resulting trajectory for a set of 6 MSs located in pairs

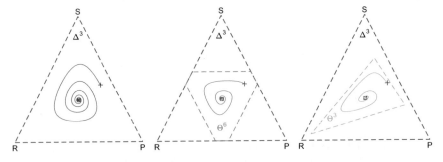

Fig. 4.2 Simplex Δ^3 in $\mathbb{R}^3$. Vertices of convex sets Θ^6 and Θ^3 represent mixed strategies $\mathbf{c}_m$. In blue, trajectory when the set of MSs is exactly the set of pure strategies. In orange and green, trajectories for the RPS game with different sets of MSs. In this examples $a = 1$ and $b = 2$

at each of the three boundaries of the simplex, as it is shown. The third trajectory is the green one, which is defined by a set of 3 MSs placed inside the simplex. Both sets of MSs are given in,

$$
C_{\Theta^6} = \begin{bmatrix} 0.55 & 0.45 & 0 & 0 & 0.45 & 0.55 \\ 0.45 & 0.55 & 0.55 & 0.45 & 0 & 0 \\ 0 & 0 & 0.45 & 0.55 & 0.55 & 0.45 \end{bmatrix}, \quad C_{\Theta^3} = \begin{bmatrix} 0.9 & 0.05 & 0.1 \\ 0.05 & 0.675 & 0.25 \\ 0.05 & 0.275 & 0.65 \end{bmatrix}.
$$

The initial distribution for pure strategies is $\mathbf{x}(0) = [0.1, 0.45, 0.45]^T$, which for the subset Θ^6 corresponds to an initial distribution for mixed strategies $\mathbf{y}(0) = [0.05, 0.05, 0.4, 0.4, 0.05, 0.05]$. On the other hand, for the subset Θ^3, it corresponds to the initial distribution for mixed strategies $\mathbf{y}(0) = [0.0304, 0.4848, 0.4848]$.

4.2.3 MSD as a Gradient Flow

The standard simplex Δ^M is a differentiable manifold also known as the *Shahshahani manifold* after the mathematician who introduced it in the field of Mathematical Biology [Har09, Leb06]. Given a vector $\mathbf{y}$ in the interior of Δ^M, the corresponding tangent space is defined as $T_{\mathbf{y}}\Delta^M = \{\mathbf{v} \in \mathbb{R}^M : \sum_{m=1}^{M} v_m = 0\}$. Let $\mathbf{a}$ and $\mathbf{b}$ be two vectors in the tangent space $T_{\mathbf{y}}\Delta^M$. For these two vectors, an alternative inner product is defined by $\langle \mathbf{a}, \mathbf{b} \rangle_{\mathbf{y}} = \sum_{m=1}^{M} a_m b_m / y_m$. It differs from the Euclidean inner product in the sense that it is defined relative to the position of $\mathbf{y}$ in the simplex, giving more weight near its boundaries.

This relative inner product is useful to define the *Shahshahani gradient*. Given a potential function $U(\mathbf{y})$, the Euclidean gradient $\nabla U(\mathbf{y})$ is uniquely defined as the vector whose euclidean inner product with a given vector $\mathbf{v} \in T_{\mathbf{y}}\Delta^M$, produces the directional derivative $D_{\mathbf{y}}U(\mathbf{v})$ of $U(\mathbf{y})$ along $\mathbf{v}$, i.e. $D_{\mathbf{y}}U(\mathbf{v}) = \mathbf{v}^T \nabla U(\mathbf{y})$. Accordingly, the Shahshahani gradient $\nabla_{sh} U(\mathbf{y})$ is defined in a similar way, with the relative inner product as $D_{\mathbf{y}}U(\mathbf{v}) = \langle \mathbf{v}, \nabla_{sh} U(\mathbf{y}) \rangle_{\mathbf{y}}$.

Given the scalar field (or potential function) $U(\mathbf{y})$, a state $\hat{\mathbf{y}}$ is defined as an Evolutionary Stable State (ESS) if it satisfies

$$
(\hat{\mathbf{y}} - \mathbf{y})^T g(\mathbf{y}) = D_{\mathbf{y}}U(\hat{\mathbf{y}} - \mathbf{y}) > 0
$$

for all $\mathbf{y} \neq \hat{\mathbf{y}}$ in a neighborhood of $\hat{\mathbf{y}}$. In other words, the ESS is a state such that a vector $\hat{\mathbf{y}} - \mathbf{y}$ and the gradient always form an acute angle, meaning that the value of the scalar field $U(\mathbf{y})$ can always be improved. Thus, $\hat{\mathbf{y}}$ is a local maximizer of $U(\mathbf{y})$. If at $t = 0$, $\mathbf{y}$ is in the neighborhood of $\hat{\mathbf{y}}$, (4.3) asymptotically converges to it [HS88, Men01, Har09].

Letting $f_k(\mathbf{x}) = \partial U / \partial x_k$ be an Euclidean gradient on $\mathbb{R}^K$ for the potential function $U(\mathbf{x})$, then the MSD Eq. (4.3) is a Shahshahani gradient with potential function $U(C\mathbf{y})$ [HS88].

To proof this, it is important to notice that (4.3) represents a vector in the tangent space $T_{\mathbf{x}}\Delta^K$ of the original simplex Δ^K since $\sum_{m=1}^{M} \dot{y}_m = 0$, and $\sum_{k=1}^{K}(\mathbf{C}\dot{\mathbf{y}})_k = \sum_{k=1}^{K} \dot{x}_k = 0$ by the definition of $\mathbf{C}$. Besides, let us use an M-dimensional vector $\mathbf{v}$ such that $\sum_{m=1}^{M} v_m = 0$. This vector produces another vector $\mathbf{z} = \mathbf{Cv}$ in the tangent space of the original simplex as well. With these inputs, the definition of the Shahshahani inner product can be applied to (4.3) as follows,

$$\langle \dot{\mathbf{y}}, \mathbf{v}\rangle_{\mathbf{y}} = \sum_{m=1}^{M} \frac{\dot{y}_m v_m}{y_m} = \sum_{m=1}^{M}((\mathbf{C}^{\mathsf{T}}\mathbf{f}(\mathbf{Cy}))_m - \bar{g}(\mathbf{y}))v_m = \sum_{m=1}^{M}(\mathbf{C}^{\mathsf{T}}\mathbf{f}(\mathbf{Cy}))_m v_m - \bar{g}(\mathbf{y})\sum_{m=1}^{M} v_m$$
$$= \mathbf{v}^{\mathsf{T}}\mathbf{C}^{\mathsf{T}}\mathbf{f}(\mathbf{Cy}) = \mathbf{z}^{\mathsf{T}}\mathbf{f}(\mathbf{Cy}) = D_{\mathbf{Cy}}U(\mathbf{z}),$$

which proofs the statement. An alternative but equivalent definition of the MSD and its proof are provided in [HS88].

4.2.4 Examples

To illustrate the concepts of the last subsection, let us consider the following example. A potential function is defined as $U(\mathbf{x}) = -(x_1 - 26/30)^2 - (x_2 - 11/30)^2 - (x_3 - 8/30)^2$, for $\mathbf{x} \in \mathbb{R}^3$. This function has a global maximum in $\hat{\mathbf{x}}^* = [26/30, 11/30, 8/30]^{\mathsf{T}}$, which is a point in the plane $\{x_1 + x_2 + x_3 = 1.5\} \in \mathbb{R}^3$. However, if the feasible region is constrained to points $\mathbf{x} \in \Delta^3 = \{x_1 \geq 0, x_2 \geq 0, x_3 \geq 0, x_1 + x_2 + x_3 = 1\}$, then the feasible maximum is $\hat{\mathbf{x}} = [7/10, 2/10, 1/10]^{\mathsf{T}}$. Both unfeasible and feasible maxima can be observed in Fig. 4.3.

In this example, it is possible to illustrate a comparison between the Euclidean gradient and the Shahshahani gradient vectors. For several points $\mathbf{x}$ in the feasible

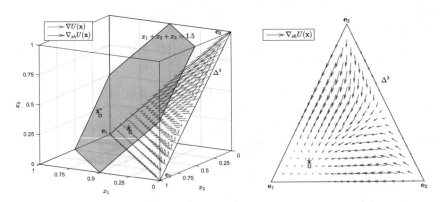

Fig. 4.3 Euclidean and Shahshahani gradient vector fields (scaled) for an example of potential function with domain $\mathbf{x} \in \mathbb{R}^3$. The feasible maximum $\hat{\mathbf{x}}$ lies in the simplex Δ^3 while the unfeasible maximum $\hat{\mathbf{x}}^*$ lies in the plane $\{x_1 + x_2 + x_3 = 1.5\} \in \mathbb{R}^3$. ($\square$: feasible and unfeasible maxima)

region Δ^3, Fig. 4.3 shows the vector fields of both gradients. As it can be observed, the Euclidean gradient vector $\nabla U(\mathbf{x})$ points to the plane $\{x_1 + x_2 + x_3 = 1.5\}$ which contains the unconstrained maximum $\hat{\mathbf{x}}^*$. This happens because the Euclidean gradient points in the direction of maximal increase of the potential function $U(\mathbf{x})$. On the other hand, the Shahshahani gradient vector $\nabla_{sh}U(\mathbf{x})$ lies on the tangent space $T_{\mathbf{x}}\Delta^3 = \{x_1 + x_2 = x_3 = 0\}$ of the simplex Δ^3, which in words means that both its tail (i.e. $\mathbf{x}$) and tip (i.e. $\mathbf{x} + \nabla_{sh}U(\mathbf{x})$) lie on the plane $\{x_1 + x_2 + x_3 = 1\} \supset \Delta^3$. On the right plot in Fig. 4.3 it is possible to clearly check the vector field of the Shahshahani gradient of the example, and the feasible maximum as well [Leb06, Har09].

Using the same potential function, let us consider a second example with a more constrained region, subset of the simplex Δ^3, defined by $\mathbf{x} \in \Theta^4 = \{0 \le x_1 \le 0.55, x_2 \ge 0, x_3 \ge 0, x_1 + x_2 + x_3 = 1\}$. This convex subset Θ^4 can also be represented by the convex hull whose vertices are the column vectors of the following MSs matrix,

$$\mathbf{C}_{\Theta^4} = \begin{bmatrix} 0.55 & 0 & 0 & 0.55 \\ 0.45 & 1 & 0 & 0 \\ 0 & 0 & 1 & 0.45 \end{bmatrix}.$$

In this second example, the feasible maximum is now $\hat{\mathbf{x}} = [22/40, 11/40, 7/40]^T$, given the new constraint $x_1 \le 0.55$. In Fig. 4.4, both the unfeasible maxima from before are labeled $\hat{\mathbf{x}}^+$, and $\hat{\mathbf{x}}^*$ respectively. The sample points $\mathbf{x} \in \Delta^3$ on Fig. 4.4 are obtained from uniformly distributed sample points $\mathbf{y} \in \Delta^4$, by applying (4.1). On Fig. 4.5 it is possible to check the original sample points $\mathbf{y} \in \Delta^4$ and the corre-

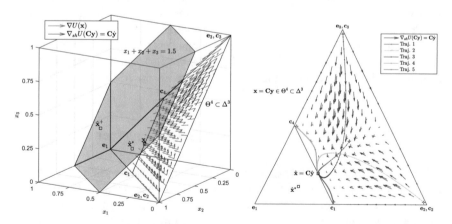

Fig. 4.4 Euclidean and Shahshahani gradient vector fields (scaled) for an example of potential function with domain $\mathbf{x} \in \mathbb{R}^3$. The feasible maximum $\hat{\mathbf{x}}$ lies in the subset Θ^4 of the simplex Δ^3, while unfeasible maxima $\hat{\mathbf{x}}^*$, and $\hat{\mathbf{x}}^+$ lie in planes $\{x_1 + x_2 + x_3 = 1\}$, and $\{x_1 + x_2 + x_3 = 1.5\}$ respectively, in $\mathbb{R}^3$. Shahshahani gradient vector field, and 5 different trajectories converging to the maximum, as seen in Δ^3. (+: initial distributions, □: maxima). © [2017] IEEE. Reprinted, with permission, from [Ova+16b]

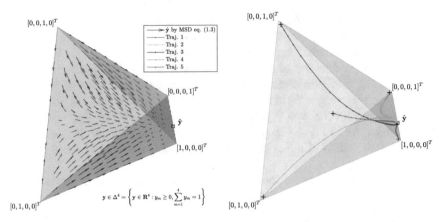

Fig. 4.5 Representation of the simplex $\Delta^4 \in \mathbb{R}^3$ as a tetrahedron in $\mathbb{R}^3$. Shahshahani gradient vector field (scaled) for an example of potential function with domain $\mathbf{x} \in \mathbb{R}^3$. The feasible maximum $\hat{\mathbf{y}}$ lies in the simplex Δ^4. 5 different trajectories converging to the maximum, as seen in Δ^4. (+: initial distributions, $\Box$: maxima). The simplex $\Delta^4 \in \mathbb{R}^4$ is a portion of the hyper-plane $\{0, y_1 + y_2 + y_3 + y_4 = 1\} \in \mathbb{R}^4$. © [2017] IEEE. Reprinted, with permission, from [Ova+16b]

sponding vector field of the Shahshahani gradient $\dot{\mathbf{y}}$ for the potential function of the example. Figure 4.4 shows a profile view of the simplex Δ^3 and the subset Θ^4. It also illustrates the Shahshahani gradient vector field $\dot{\mathbf{y}}$ within the convex hull given by the MSs of $\mathbf{C}_{\Theta^4}$, i.e. $\dot{\mathbf{x}} = \mathbf{C}\dot{\mathbf{y}}$.

In Fig. 4.5 it is possible to check 5 different trajectories, for the evolution of the population, starting from different initial distributions $\mathbf{y}$ within Δ^4. The equivalent trajectories in Δ^3 obtained with (4.1), are plotted on Fig. 4.4 as well. It is possible to check that trajectories converge to the feasible optimal.

These features of the MSD motivate the development of the proposed approach for decentralized management of PEV load. In the next section, details on the established analogies and the definitions of the payoff functions are provided.

4.3   The MSD Approach for PEV Load Management

Taking into account the MSD description and following some analogies, a multi-population model is proposed for the PEV load scheduling problem. In this multi-population model, energetic quantities must be allocated to *time slots* according to a certain benefit. First, the total load of a transformer in a distribution system is considered as the *total population*. The forecast load of the transformer is represented as a *sedentary population* that does not follow any dynamics. On the other hand, the numerous controllable PEV loads are represented by *nomad populations* choosing among mixed strategies (defined by PEV owners as it will be observed). Each time slot considered in the forecast horizon is represented as an environment or pure strategy.

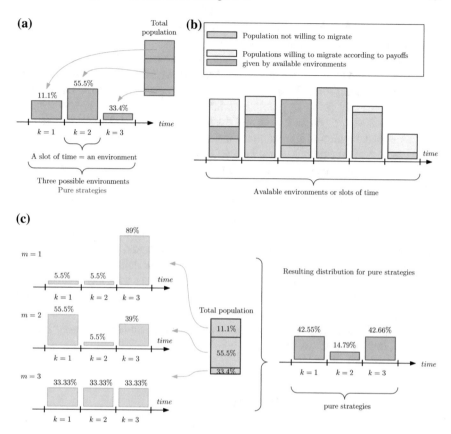

Fig. 4.6 a Example of a distribution of a single population (PEV load) in three environments (pure strategies, or time slots). **b** Example of a multi-population model, with populations willing to migrate (PEV load) through available environments, and sedentary populations (Base load forecast). **c** Example of distribution of a single population (PEV load) among three predefined MSs, and the resulting distribution among pure strategies. © [2017] IEEE. Reprinted, with permission, from [Ova+16b]

Each nomad population can be distributed in several time slots or environments. The distribution of single populations will evolve following the MSD according to the payoff obtained from their mixed strategies. As it is defined before, MSs are convex combinations of pure strategies. In this case a mixed strategy can be seen as one of multiple options for fully charging the PEV. Conversely, MSs are ways of distributing the individual population among environments (among time slots).

Examples of these analogies are illustrated in Fig. 4.6. In Fig. 4.6a, pure strategies are all the possible time slots (hour steps) where PEVs can allocate its energy consumption to fully charge their batteries. Time slots may have previously allocated load that corresponds to the forecast load of the transformer. In this multi-population scenario, load from PEVs and load from the forecast are both interpreted as nomad and sedentary populations respectively. An example illustrates these nomad and sedentary population analogies in Fig. 4.6b.

The role of MSs is illustrated in Fig. 4.6c. In this example the population of one PEV is distributed among three predefined MSs. These MSs are predefined possible ways of distributing the PEV load in the available environments, for instance the first mixed strategy ($m = 1$) pre-defines 5.5% of the population allocated in the environment $k = 1, 5.5\%$ on $k = 2$, and the 89% left on $k = 3$. Taking into account the pre-defined population allocations of the other two MSs, if the PEV population is distributed among the three MSs as 11.1% for $m = 1$, 55.5% for $m = 2$, and 33.4% for $m = 3$, then the resulting load distribution obtained with (4.1) is 42.55% for $k = 1$, 14.79% for $k = 2$, and 42.66% for $k = 3$.

Nomad populations make evolve their distributions in time according to the payoffs they get from the environments. The payoff seen by a portion of a single population established in a time slot (a pure strategy) is defined by a global measure and a local measure. The global measure is established based on the entropy of the global distribution of the population. The local measure is defined by the entropy of the single population. The payoff function is established as a trade-off between these two measures. This trade-off is different for each individual population and depends on the PEV owner. She/He defines the trade-off based on the incentive given by the utility grid manager. The payoff corresponding to MSs depend on the payoff for pure strategies, and on the set of MSs itself.

4.3.1 Payoff Functions Definition

Let us consider the vector $\mathbf{l} = [l_1, l_2, \ldots, l_k, \ldots, l_K]^\mathrm{T}$ as the forecast for the load profile of a transformer in a distribution system (*sedentary* population). Also, let $\mathbf{x}^j = [x_1^j, x_2^j, \ldots, x_k^j, \ldots, x_K^j]^\mathrm{T}$ be the power consumption profile for a PEV with index $j = \{1, \ldots, J\}$ (a single *nomad* population). The total load (global population) in K periods is a scalar,

$$\mu = \sum_{k=1}^{K} \left(l_k + \sum_{j=1}^{J} x_k^j \right),$$

and the distribution of this load over the whole window of time is represented by $\lambda = [\lambda_1, \lambda_2, \ldots, \lambda_k, \ldots, \lambda_K]^\mathrm{T}$, where the portion of load assigned to period k is $\lambda_k = l_k + \sum_{j=1}^{J} x_k^j$.

Adopting the concept from information theory [KK89], the Shannon's entropy of this load distribution is defined as,

$$S(\lambda) = -\sum_{k=1}^{K} \lambda_k \ln\left(\frac{\lambda_k}{\mu}\right) = -\sum_{k=1}^{K} \left(l_k + \sum_{j=1}^{J} x_k^j \right) \ln\left(\frac{l_k}{\mu} + \sum_{j=1}^{J} \frac{x_k^j}{\mu}\right), \quad (4.4)$$

and it should be noticed that $\lambda_k/\mu \geq 0$, and $\sum_{k=1}^{K} \lambda_k/\mu = 1$. Then, it is possible to act over the distribution to affect its entropy. More specifically, each PEV i is able to modify its local load distribution $\mathbf{x}^i$ to affect the entropy of the total load distribution. In fact, local distributions $\mathbf{x}^i$ have to be such that the battery is fully charged at the end of the charging periods. Then the size of local populations is defined by,

$$\Gamma^i = \frac{soc_d^i - soc_0^i}{\tau},$$

where soc_d^i, and soc_0^i are the desired and initial states of charge of the PEV i respectively, and τ is the time length of each slot k. Based on this, the Shannon's entropy of the distribution of a single population $\mathbf{x}^i$ is defined by,

$$S(\mathbf{x}^i) = -\sum_{k=1}^{K^i} x_k^i \ln\left(\frac{x_k^i}{\Gamma^i}\right) \qquad (4.5)$$

where K^i is the number of time slots allowed for population i. This value is defined by the owner who can be as restrictive as she/he wants with the time (respecting the charger's constraints and eventually some level of exigency from the utility grid manager). It also should be noticed that $x_k^i/\Gamma^i \geq 0$, and $\sum_{k=1}^{K^i} x_k^i/\Gamma^i = 1$.

The Shannon's entropy is a strictly concave function on parameters lying on the standard simplex [KK89]. By now, it should be noticed that the total load distribution lies in the simplex

$$\Delta_\mu^K = \left\{\lambda \in \mathbb{R}^K : \lambda_k \geq 0, \sum_{k=1}^{K} \lambda_k = \mu\right\}, \qquad (4.6)$$

and the local load distributions lie in simplices

$$\Delta_{\Gamma^i}^{K^i} = \left\{\mathbf{x}^i \in \mathbb{R}^{K^i} : x_k^i \geq 0, \sum_{k=1}^{K^i} x_k^i = \Gamma^i\right\}. \qquad (4.7)$$

To proof that the considered entropy functions are maximized if all the arguments on their corresponding natural logarithms are equal, it is possible to apply the inequality of the weighted geometric and arithmetic means (WGM and WAM) [KK89], as follows. For the case of the total load distribution (4.4), it should be noticed that $\exp(S) = \prod_{k=1}^{K}(\mu/\lambda_k)^{\lambda_k}$ is the WGM (to the power of μ) of K elements μ/λ_k with K non-negative weights λ_k. Then, the inequality is given by,

$$\prod_{k=1}^{K}\left(\frac{\mu}{\lambda_k}\right)^{\lambda_k} \leq \left(\frac{\sum_{k=1}^{K}(\mu/\lambda_k)\lambda_k}{\sum_{k=1}^{K} \lambda_k}\right)^\mu = \left(\frac{\mu K}{\mu}\right)^\mu = K^\mu,$$

which holds for equality if and only if the K considered elements are equal, giving maximum value $\hat{S} = \ln(\exp(\hat{S})) = \ln(K^{\mu}) = \mu \ln(K)$. The same proof sketch can be applied to the local case (4.5). From this, let us define payoff functions

$$
f_k^i(x_k^i) = -\alpha^i \ln \left(\frac{l_k}{\mu} + \sum_{j=1}^{J} \frac{x_k^j}{\mu} \right) - (1 - \alpha^i) \ln \left(\frac{x_k^i}{\Gamma^i} \right) - 1,
$$

where α^i is the trade-off factor between the utility grid manager's interests and the owner's interests who looks for preserving its battery by preventing large variations on the charging rate over time. In principle $0 \le \alpha^i \le 1$, which means that an owner choosing $\alpha^i \rightarrow 1$ is very interested in the utility's offered incentive. Furthermore, this trade-off factor induces fairness in the participation of PEVs in the entropy maximization of the total load distribution. This will be checked in the results section.

If PEV i has a trade-off factor close unity, it is likely that its dynamics will assign to a slot k, a larger portion of its population. In this scenario, the MSs approach is a viable solution to introduce the constraints on the chargers, as it will be introduced in the next section.

4.3.2 Representing Charger's Constraints with Mixed Strategies

If the payoff obtained from a pure strategy is very high compared to other pure strategies, then the MSD will tend to increase the portion of the population following that strategy. In the PEV case this is not viable since chargers are physically constrained in terms of nominal power limits. Thus the load distribution has to be constrained to the following subset Θ^{M^i} of the simplex $\Delta_{\Gamma^i}^{K^i}$, or even a subset of Θ^{M^i} itself,

$$
\Theta^{M^i} = \left\{ \mathbf{x}^i \in \mathbb{R}^{K^i} : 0 \le x_k^i \le \overline{p}^i, \sum_{k=1}^{K^i} x_k^i = \Gamma^i \right\},
$$

where $\overline{p}^i$ is the predefined upper limit of the charger of PEV i. In this scenario, the total population can be divided at most in γ^i pure strategies, where

$$
\gamma^i = 1 + \left\lfloor \frac{\Gamma^i}{\overline{p}^i} \right\rfloor,
$$

and $\lfloor \cdot \rfloor$ refers to the largest previous integer. As it can be seen, γ^i can be at least 1 and at most K^i. The MSs are all the permutations of a vector of K^i elements where:

- the first $(\gamma^i - 1)$ elements are equal to $\overline{p}^i / \Gamma^i$,
- the next element is equal to $(1 - \overline{p}^i (\gamma^i - 1)/\gamma^i)$,

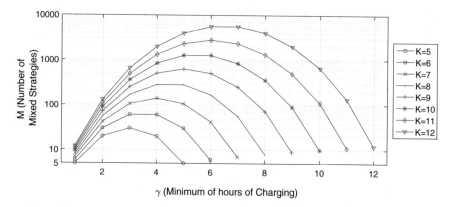

Fig. 4.7 The number of mixed strategies M described by (4.8), as function of number of pure strategies K, and the minimum amount γ of hours of charge. © [2017] IEEE. Reprinted, with permission, from [Ova+16a]

- the other $(K^i - \gamma^i)$ elements are equal to 0.

These permutations are vectors whose elements are positive and sum 1 as it is expected for the MSs (in a normalized set). Thus, the total number of MSs needed to span Θ^{M^i} in the convex combination is

$$M^i = \frac{K^i!}{(\gamma^i - 1)!(K^i - \gamma^i)!}. \tag{4.8}$$

As it can be inferred from (4.8), even if the number of pure strategies is relatively high, if γ^i is relatively close to 1 or to K^i, then M^i is not very large and finding the full set of MSs is not computationally expensive. The relation among γ^i, K^i and M^i can be checked in Fig. 4.7. The whole set of MSs can be enumerated by some recursive algorithm. For the state vector $\mathbf{x}^i$, let us define a MSD state vector $\mathbf{y}^i$ such that $\mathbf{x}^i = \mathbf{C}^i \mathbf{y}^i$ as it is described by (4.1). Elements y_m^i behave according to the MSD Eq. (4.3). This state vector represents the parameters for payoff functions for mixed strategies $\mathbf{g}^i(\mathbf{y}^i)$ as it is defined by (4.2), $\mathbf{g}^i(\mathbf{y}^i) = \mathbf{C}^{iT}\mathbf{f}^i(\mathbf{C}^i\mathbf{y}^i)$, where payoffs for pure strategies $\mathbf{f}^i(\cdot)$ are defined by the proposed payoff functions.

4.4 The MSD Based Distributed Management Loop

Based on the details of the proposed approach, the discretization of (4.3) results in the distributed algorithm presented on Fig. 4.8. In this section, a description of the algorithm is presented. Two parties are involved, PEVs which act separately and independently over their load distributions, and an aggregator in charge of collecting, aggregating, updating, and redistributing information to and from PEVs.

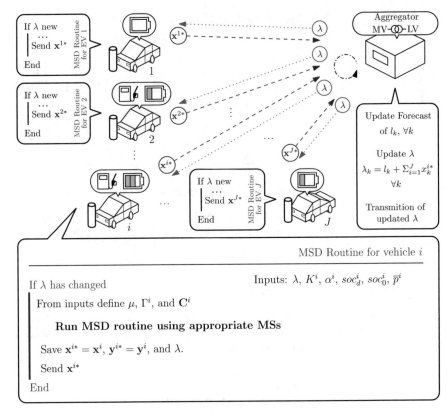

Fig. 4.8 Diagram describing the decentralized MSD approach

4.4.1 The Role of PEVs in the Loop

Each PEV, equipped with a local MSD algorithm, locally manipulates its consumption rate profile $\mathbf{x}^i$, having previously defined its owner's trade-off parameter α^i. The number of time slots K^i (number of pure strategies) corresponding to each PEV depend also on the owner. The PEV initially defines its local load distribution uniformly, without optimizing it, i.e. $x_k^i = (soc_d^i - soc_0^i)/K^i$, $\forall k$.

The parameter K^i is redefined with the evolution of time, getting smaller until the expected time of departure, defined by the owner as well. The local evolution given by the MSD runs with the computed MSs for the present value of K^i, inputs soc_d^i, soc_0^i, α^i, and the nominal power of the charger $\overline{p}^i$. Another input to the MSD is the total load distribution λ obtained from the aggregator.

The number of MSs reduces with the evolution of time since K^i becomes smaller as the programmed disconnection time approaches. When an updated total load distribution λ is received from the aggregator, the previous $\mathbf{x}^{i*}$ is subtracted from the new λ, and the MSD routine is executed again to find the new local $\mathbf{x}^{i*}$. Every new execution of the routine begins with the previously stored distribution of the local population over MSs. Once the algorithm reaches the limiting amount of iterations or fulfills the convergence criterion, the $\mathbf{x}^{i*}$ is sent to the aggregator. The interaction among local populations is based on the BR dynamics [HS88], applicable since the proposed approach considers a nonnegative weighted sum of the global and local entropies, concave functions on their domains (4.6) and (4.7), which is concave as well [BV04].

It is considered that the evolution has reached an ESS when the weighted variance of payoff functions $\mathbf{g}^i(\mathbf{y}^i)$, goes below a predefined tolerance [Har09]. This stopping criterion, mentioned on Fig. 4.8, is established since the rate of change of the potential function, $\dot{U}(\cdot)$, is its directional derivative along $\dot{\mathbf{x}} = \mathbf{C}\dot{\mathbf{y}}$, given by,

$$D_{\mathbf{C}\mathbf{y}}U(\mathbf{C}\dot{\mathbf{y}}) = \langle \dot{\mathbf{y}}, \dot{\mathbf{y}} \rangle_{\mathbf{y}} = \sum_{m=1}^{M} y_m ((\mathbf{C}^{\mathrm{T}}\mathbf{f}(\mathbf{C}\mathbf{y}))_m - \bar{g}(\mathbf{y}))^2.$$

The local MSD routine is provided as two Matlab scripts in Sect. 4.6 at the end of this chapter. It includes step by step descriptions aiming to help the reader to understand how it behaves. It is also written and described to help the reader to re-program it other platforms according to her/his needs.

4.4.2 The Role of the Aggregator in the Loop

An aggregator is in charge of managing the interaction among populations on a BR dynamics. It collects the population profiles $\mathbf{x}^i$ of each PEV i. It is in charge of providing the short-term load forecast $\mathbf{l}$ for the MV-LV transformer of the distribution system, based on proper historic data of load, weather and seasonal information. The aggregator compiles information from the PEVs and the forecast, aggregates it as $\lambda_k = l_k + \sum_{i=1}^{J} x_k^{i*}$, $\forall k = \{1, \ldots, K\}$, and redistributes the new profile with a communication mechanism to reach all the PEVs under the transformer of the distribution system. It is constantly receiving and sending updated information to each of the PEVs. In principle, when new information arrives, the aggregator updates and continues with the distribution of information.

For a PEV, fulfilling the variance criterion for local convergence is not as important as, for a reduced amount of iterations, moving towards a better distribution $\mathbf{x}^{i*}$ in terms of the payoff functions, even if the reached distribution is still suboptimal. This is because at each local iteration, other PEVs are not able to change their corresponding distributions. Thus, if a local ESS is reached in the current round of iterations, it is likely to be ignored and changed in the next round of iterations because it will be suboptimal (given that other PEVs will also change their distributions). This is verified in the results section.

4.5 Numerical Examples

In this section, the proposed method is tested and the results are separated in two subsections. The first subsection studies in detail the proposed MSD multi-population method. Besides, this subsection is proposed to study the approach when full sets of MSs are employed by each PEV in its local MSD routine. On the other hand, the second subsection is intended to illustrate how the method works when reduced sets of the most convenient MSs are employed. This second subsection describes the behavior of the method when MSs are chosen to privilege early high (or fast) charging rates, but still maintaining fairness, and grid impact reduction.

4.5.1 Using Complete Sets of Mixed Strategies

The proposed approach is tested using real active power consumption measured from a distribution transformer for residential purposes and sport facilities, from the SOREA utility grid company [SOR] in the region of Savoie, France. A first dummy scenario is presented in Fig. 4.9, to illustrate the effect of trade-off factors α^i. In this case, 30 PEVs sharing the same energy needs (6 kWh), power limits (3 kW), and trade-off factors $\alpha^i = \alpha$, are connected during 24 h. A summary of these assumptions can be found on Table 4.1. Figure 4.9 shows the final distributions of the total load including the forecast of the base load and the 30 PEVs, obtained by the proposed approach for three values of the trade-off factor α. It is possible to observe that, for the utility grid manager's interests, it is desirable to have trade-off factors α^i as close to unity as possible, to allocate PEVs' load to off-peak periods of time.

To illustrate the effect of the definition of MSs as suggested in this chapter, let us consider another dummy scenario shown in Fig. 4.10. This time with only 5 PEVs sharing the same energy needs (8 kWh), power limits (3 kW), and identical trade-off factors $\alpha^i = 1$, connected during 10 night hours from 21 to 07 h in a similar day as the one presented in the scenario before. A summary of these assumptions can be found on Table 4.2. Diagrams on the top show a representation of the evolution of local load distributions obtained with the proposed approach in two cases of the same scenario. Each vertex of the polygons represents a pure strategy (one hour in the ten hour period), and the trajectories shown within the polygons are the convex combinations of the vertices with coefficients given by the state vectors $\mathbf{x}^i$ evolving with the MSD in both cases. The barycenters of the polygons represent uniform distributions, where all the local load distributions start. Each PEV updates its load distribution asynchronously. The four charts on the left in Fig. 4.10 show the case where the set of MSs for each PEV is the identity matrix. It can be observed that without the proper definition of MSs, local distributions are able to reach states forbidden by the charger. For instance, the first vehicle participating in the algorithm assigns all of its load between 23 h and 00 h, mostly at 00 h where its consumption is close to 7 kW, more than the double of the charger's limit.

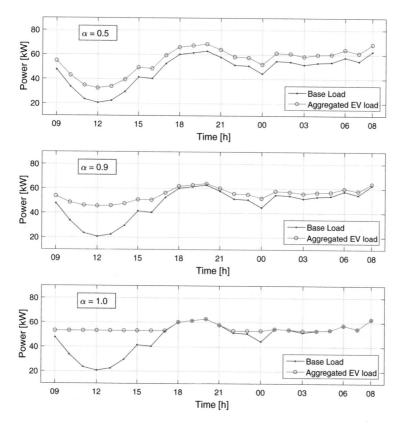

Fig. 4.9 Example of 30 PEVs connected during 24 h, all sharing the same energy needs (6 kWh), charger limits (3 kW), and trade-off factor α in three scenarios ($\alpha = 0.5$, $\alpha = 0.9$, and $\alpha = 1.0$). Resulting total load distributions. © [2017] IEEE. Reprinted, with permission, from [Ova+16a]

Table 4.1 Descriptive summary and assumptions of study case on Fig. 4.9

Item	Description
Number of considered PEVs	30 PEVs
Chargers	All of them 3 kW
Energy requirements	All of them 6 kWh
Trade-off factors α^i	Common to all $\alpha^i = \alpha$
Distribution system info.	Data from SOREA utility grid company [SOR]
Evaluated scenarios	Variation on the values of the common trade-off factor α
Time period	24 h (hour steps)

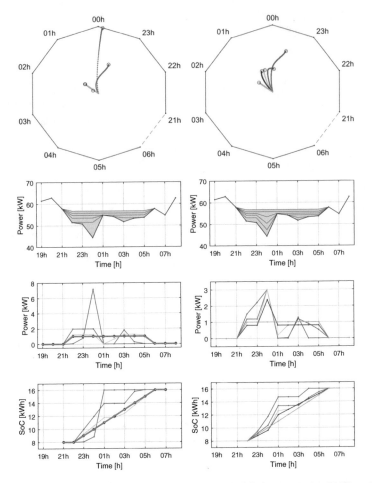

Fig. 4.10 Example of 5 PEVs sharing same energy needs (8 kWh), power limits (3 kW), and trade-off factors ($\alpha = 1$). Left: a case where MSs are exactly the set of pure strategies. Right: a case where MSs are defined like in Sect. 4.3.2. © [2017] IEEE. Reprinted, with permission, from [Ova+16a]

On the other hand, diagrams on the right of Fig. 4.10 show the case where the set of MSs for each PEV is defined like in Sect. 4.3.2. In this case, the assignation of load to each hour is limited to be in the convex hull defined by MSs. In this case the first PEV in the interaction is not allowed to assign all of its consumption between 23 and 00 h, thus it is forced to assign load to the next desirable hours in terms of entropy maximization (22 and 03 h).

In the last example it is possible to notice that only one update per PEV is needed to converge to the best possible total load distribution, given that $\alpha = 1$. In fact, more updates will not change the reached local load distributions of each PEV. Figure 4.11 shows the corresponding evolution in terms of entropy maximization contributed by the 5 PEVs, given that all trade-off factors are $\alpha = 1$. In this diagram, entropy is

Table 4.2 Descriptive summary of study cases on Figs. 4.10 and 4.13

Item	Description
Number of considered PEVs	5 PEVs
Chargers	All of them 3 kW
Energy requirements	All of them 8 kWh
Trade-off factors α^i	Common to all $\alpha^i = \alpha$
Distribution system info.	Data from SOREA utility grid company [SOR]
Evaluated scenarios	Variation on the values of the common trade-off factor α $\alpha = 1$, $\alpha = 0.98$, and $\alpha = 0.9$
Time period	10 h steps ($K^i = 10$) between 21 and 07 h (disconnection at 07h)

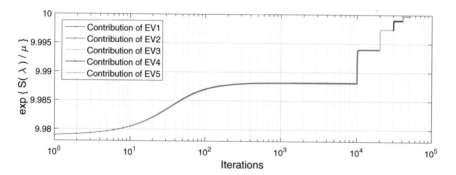

Fig. 4.11 Evolution of the total load distribution entropy with the participation of PEVs in the proposed algorithm, for the example of 5 PEVs connected, sharing the same energy needs (8 kWh), power limits (3 kW), trade-off factors ($\alpha = 1$) and MSs defined like in Sect. 4.3.2. © [2017] IEEE. Reprinted, with permission, from [Ova+16a]

plotted as $\exp(S(\lambda)/\mu)$ which is expected to approach to $K^i = K = 10$, since the defined total load distribution entropy $S(\lambda)$ as defined in Sect. 4.3.1, is expected to converge to $\mu \ln(K)$ and μ is the total load of the studied period. It can be noticed that the final total distributions are as uniform as possible, thus the proposed objectives achieve valley filling. For the first PEV in this example, the evolution of the weighted variance of payoff functions, the distributions by MSs and by pure strategies are shown in Fig. 4.12 (360 MSs given the 3 kW power limit, the charging period length $K = 10$, and the 8 kWh energy need).

However, even if convergence to the maximum entropy of the total load distribution is fast for $\alpha = 1$, there is an unfair allocation of resources as it can be checked on the evolution of the states of charge in Fig. 4.10. Here some PEVs (blue and red in this case) are able to charge their PEVs in more favorable conditions depending on their position in the interaction (faster in this scenario for the first PEVs). In this sense, the trade-off factors introduce fairness in the allocation of resources among PEVs.

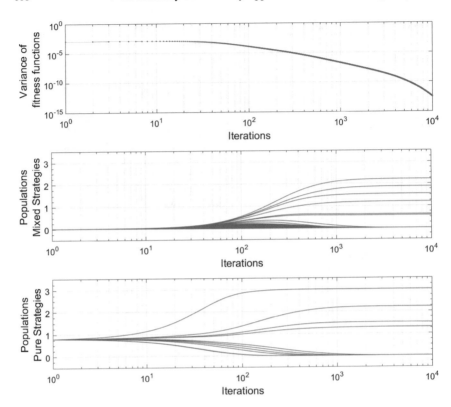

Fig. 4.12 Evolution of the local MSD routine for the first PEV in the 5 PEVs example. Evolution of: (*top*) weighted variance of payoff functions for MSs; (*middle*) the state vector of the distribution by MSs ($M = 360$); (*bottom*) the state vector $\mathbf{x}^i$ of the distribution by pure strategies ($K = 10$). © [2017] IEEE. Reprinted, with permission, from [Ova+16a]

Figure 4.13 shows some results for the same dummy scenario, this time considering the same MSs definition and a trade-off factor $\alpha = 0.9$ common to all PEVs. In this case, it is possible to check that each time PEVs run their local MSD routines, their local load distributions are updated and tend to asymptotically converge to a common distribution. In this case, each PEV updates its local distribution 5 times and the resulting state of charge profiles become very close for all PEVs. The resulting state of charge profiles become also very similar for all PEVs and more uniform meaning that the rates of charge at each time step are more uniform. However, as expected, this fairness improvement reduces the valley filling capability of the approach.

4.5.1.1 Example Under Realistic Conditions

In the following study case, around 40 PEVs per day are connected during two days. Battery capacities of 5 kWh (Hybrid PEVs), 20 kWh, and 30 kWh are considered to be

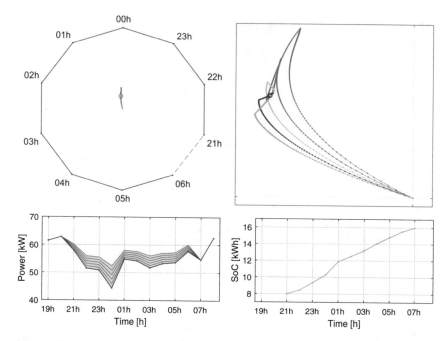

Fig. 4.13 The same example with 5 PEVs sharing the same energy needs (8 kWh), power limits (3 kW), and trade-off factors ($\alpha = 0.9$). In this case, each local distribution is updated 5 times. The five state vectors asymptotically converge to the same distribution. The upper right plot is a zoom to the trajectories. © [2017] IEEE. Reprinted, with permission, from [Ova+16a]

charged in the grid with probabilities of 30%, 50%, and 20% respectively. Considered charger's power limits are 3, 7.5, and 10 kW, with penetrations of 60%, 30%, and 10% respectively. Moreover, a high presence of PEVs is registered during peak demand hours. A summary on the assumptions of this study case can be found on Table 4.3. The number of connected PEVs and the resulting total load distributions without management can be observed in Fig. 4.14. In this figure, the total load distribution resulting from the proposed approach can be observed when $\alpha^i = 1$ for all PEVs.

Now, let us consider that the utility grid manager is able to motivate owners to use trade-off values $\alpha^i \leq 1$ with normal distribution of mean 0.8 and standard deviation 0.05. With this assumption, the corresponding total load distribution profiles obtained with the proposed approach are presented in Fig. 4.15. Additionally, the resulting state of charge profiles are presented on Fig. 4.16. Similarity in charging rates among PEVs reflects *fairness* in the allocation of resources.

Table 4.3 Descriptive summary and assumptions of the considered realistic scenario

Item	Description
Chargers	3 kW with probability of 60%
	7.5 kW with probability of 30%
	10 kW with probability of 10%
Batteries	5 kWh with probability of 30%
	20 kWh with probability of 50%
	30 kWh with probability of 20%
Constraints on batteries	Between 30 and 80% for all of them
Time period	2 and a half days (59 h steps)
Distribution system info.	Data from SOREA utility grid company [SOR]
Evaluated scenario	High presence of PEV during peak demand hours
Peak of connected PEVs	Between 27 and 30 PEVs at 20 h each day
Trade-off values α^i	Normally distributed with mean value 0.8 and standard deviation 0.05 ($\alpha^i \leq 1$)

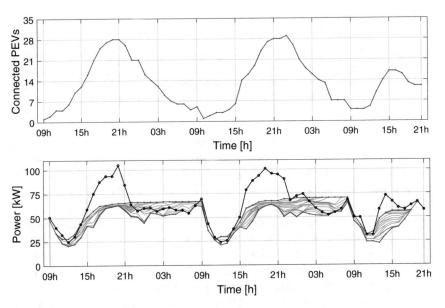

Fig. 4.14 (*top*) Connected PEVs per hour (PEVs present mostly during peak demand hours). (*bottom*) Corresponding total load distributions without management (black) and with management with $\alpha^i = 1$ (red). © [2017] IEEE. Reprinted, with permission, from [Ova+16a]

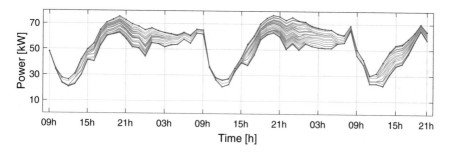

Fig. 4.15 Resulting total load distribution with random α^i values

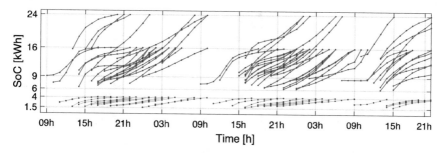

Fig. 4.16 Resulting state of charge profiles for case with random α^i values. 5 kWh batteries (yellow), 20 kWh batteries (blue), and 30 kWh batteries (red). © [2017] IEEE. Reprinted, with permission, from [Ova+16a]

4.5.2 Using Reduced Sets of Convenient Mixed Strategies

MSs are defined based on the charging rate constraints imposed by the chargers. Even if it may be desirable to find all the set of MSs, it is also possible to use those which are more interesting from the owner's perspective. For instance, owners can choose only those MSs resulting in faster PEV recharge, and the MSD will find the most convenient convex combination of them (from the entropies trade-off point of view). An example of a reduced set of MSs is shown on Fig. 4.17, for a PEV with 8 kWh energy need, 3 kW charger limit, and a charging window of length $K^i = 6$. In this case, the minimum amount of environments is $\gamma^i = 3$, and the total amount of MSs that represent completely the feasible region is $M^i = 60$, given by Eq. (4.8). Only 7 out of 60 MSs are shown on Fig. 4.17. These are the MSs allowing the fastest charging, and among these, earlier high charging rates are allowed by those marked with a square. It is important to mention that rather than specifying a general procedure for choosing the most convenient MSs, this chapter illustrates the general idea. We encourage the reader to propose an approach and apply it using the scripts included at the end of this chapter.

In the first part of this subsection, 5 PEVs are considered to be charged in a residential grid between 11 h and 17 h. These PEVs are connected to chargers limited

Fig. 4.17 Another example of a reduced set of convenient MSs, for a PEV with an energy demand of 8 kWh, $K^i = 6$ h, and a 3 kW charger. The full set has $M^i = 60$ MSs, while the reduced set has only 7 MSs. (□: the most convenient MSs privileging early high charging rates). © [2017] IEEE. Reprinted, with permission, from [Ova+16b]

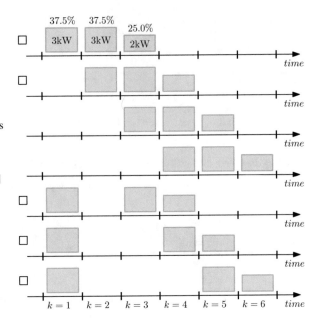

Table 4.4 Descriptive summary and assumptions of study cases on Figs. 4.18 and 4.19

Item	Description
Number of considered PEVs	5 PEVs
Chargers	All of them 3 kW
Energy requirements	All of them 8 kWh
Trade-off factors α^i	Common to all $\alpha^i = \alpha = 0.99$
Distribution system info.	Data from SOREA utility grid company [SOR]
Evaluated scenarios	Variation on the quantity of considered MSs ($M^i = 60$, $M^i = 7$, and $M^i = 4$)
Time period	6 h steps ($K^i = 6$) between 11 and 17 h (disconnection at 17 h)

in nominal power to 3 kW. In this conditions, they have 6 h to consume 8 kWh to reach a desirable state of charge (initially they have 40% and they have to reach 80% of their storage capacities of 20 kWh). Thus the total amount of MSs for each PEV is $M^i = 60$ given by (4.8), and the sets for each PEV are identical. A summary of the assumptions of this study case can be found on Table 4.4. For this case, the most convenient MSs are also those shown on Fig. 4.17.

In Fig. 4.18, it is possible to observe the final load distributions obtained by the MSD approach when all the 60 MSs are used by each PEV, when only the most convenient are used, and when only 4 out of those privileging early high rates are employed (the first two and the last two). In this subsection the trade-off factor α is fixed for all the PEVs to $\alpha = 0.99$.

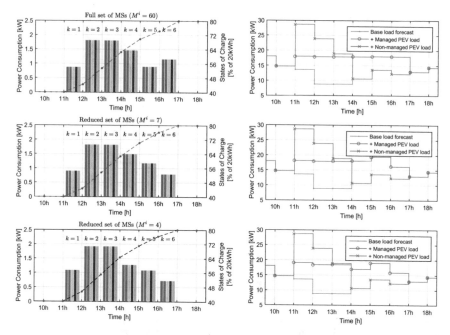

Fig. 4.18 Example of 5 PEVs, all of them with energy demands of 8 kWh, $K^i = 6$ h to charge their batteries, and 3 kW chargers. (*top*) All of them using full sets of $M^i = 60$. Center: All of them using reduced sets of 7 convenient MSs out of 60. (*bottom*) All of them using reduced sets of the 4 most convenient MSs, privileging early high charging rates. (*left*) final PEV load distributions and SoC profiles of the 5 PEVs, for the three cases. (*right*) final total load distributions, for the three cases, compared to the non-managed case. © [2017] IEEE. Reprinted, with permission, from [Ova+16b]

In these three cases, the final local load distributions are equal for all the PEVs. As it can be expected, using reduced sets of MSs still maintains the fairness property of the MSD approach, for the allocation of resources. However, when full sets are employed, the final total load distribution is almost flat, while for the scenarios with reduced sets, the final reached total load is less even. This occurs because the local load distributions evolve such that they try to reach a maximum total load distribution entropy (a total load profile as flat as possible) with the convex combination of only the available MSs. On the other hand, it can be observed that with the reduced sets, earlier high charging rates are privileged, specially for the third case where the first three hours show increased charging rates compared to the other cases.

Figure 4.19 shows the evolution of the convex combinations that define the load distributions among the 6 h for the 5 PEVs and for the three scenarios. Furthermore, recalling that MSs are convex combinations as well, Fig. 4.19 is also useful to show where the sets of MSs map inside the convex hull of the vertices representing each pure strategy (6 pure strategies). It is also possible to check how the 5 load distributions evolve and converge to the same distribution.

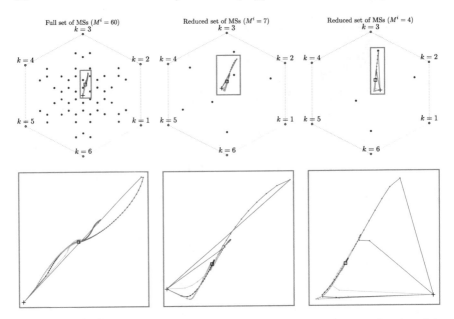

Fig. 4.19 Representation of the evolution of the 5 distributions as a convex combination of the vertices of an hexagon (6 pure strategies). Red dots represent the MSs used in the respective case. (*left*) Full set of MSs; (*middle*) Reduced set of convenient MSs; (*right*) reduced set of the 4 most convenient MSs. Plots on the bottom show zooms on the trajectories. © [2017] IEEE. Reprinted, with permission, from [Ova+16b]

On the other hand, a realistic scenario is presented in Fig. 4.21, using real historical active power consumption, measured on a distribution transformer from the SOREA utility grid company [SOR] in the region of Savoie, France. These measurements correspond to mostly office buildings. In this case, the arrival and departure of several PEVs is modeled in a random way, using a *Poisson* process model with variable rate of arrivals (changing with time) and variable connection times [Kin92]. The highest rate of arrivals is 5 PEVs/h around 05 h and it decays up to 0.5EVs/h at 04 h the next day. Vehicle battery capacities are defined to be 20 kWh. Chargers are randomly chosen to have limits of power of 3.3 kW with a probability of 80%, and 7.5 kW with a probability of 20%. A summary on these assumptions can be found on Table 4.5.

Figure 4.20 shows the number of arrivals per hour and the amount of connected PEVs per hour for the proposed test case. It should be noticed that this is a difficult scenario for the distribution transformer since the peak hours of PEV load without management coincide with the peak hours of the forecast base load of the transformer, as it can be observed in Fig. 4.21. Furthermore, Fig. 4.21 shows the most interesting results for the total load distribution using full sets of MSs, reduced sets of convenient MSs, and reduced sets of the most convenient MSs privileging early high charging rates as those of Fig. 4.17. Figure 4.21 compares the results of the proposed approach, against those obtained without any management strategy.

Table 4.5 Descriptive summary and assumptions of the considered realistic scenarios

Item	Description
Chargers	3.3 kW with probability of 80%
	7.5 kW with probability of 20%
Batteries	20 kWh for all of them
Constraints on batteries	Between 30 and 80% for all of them
Time period	4 days (96 h steps)
Highest rate of arrivals	5 PEVs/h at 05 h
Lowest rate of arrivals	0.5 PEVs/h at 04 h next day
Peak of connected PEVs	Between 24 and 32 PEVs around 11 h each day
Trade-off values α^i	Common and fixed for all PEVs to $\alpha = 0.95$
Distribution system info.	Data from SOREA utility grid company [SOR]
Evaluated scenarios	All MSs, Best MSs, and Reduced Best MSs

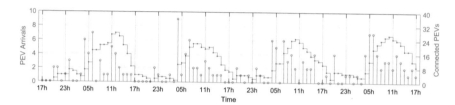

Fig. 4.20 Profiles of arrivals and connected PEVs throughout the evaluation time. The peaks of connected PEVs coincide with the peaks of the forecast load, which makes a PEV load management approach desirable. © [2017] IEEE. Reprinted, with permission, from [Ova+16b]

Despite the degradation observed in the final total load distributions with the reduction of the amount of MSs, the MSD approach, using only convenient MSs, is still able to handle PEVs' load properly. Fairness, PEV owners' convenience, and impact reduction are still achieved. Taking into account that entropy of the total load distribution is given by Eq. (4.4), entropy measurements were taken for each hour step of the first three days of the scenario on Fig. 4.21. Using time horizons of $K = 24$ hours, the results are presented on Fig. 4.23 for the ideal case (uniform total load distribution), the unmanaged case, and the managed case using the MSD approach in three modes (full sets of MSs, Best MSs, and Reduced Best MSs). It is possible to observe that in terms of total load distribution entropy maximization, MSD with full sets of MSs achieves the closest entropy measurements to the ideal case (uniform distribution). It can be observed as well that the worst measurements correspond to the unmanaged case, as it is expected. The key element to highlight is the closeness of the entropy measurements for the MSD approach using full and reduced sets of MSs, proving that the the proposed MSD approach with only convenient MSs, is still able to handle PEV load properly.

Figure 4.22 shows the corresponding state of charge profiles for the load profiles of Fig. 4.21. It should be mentioned that to reduce the impact on batteries, states

(a)

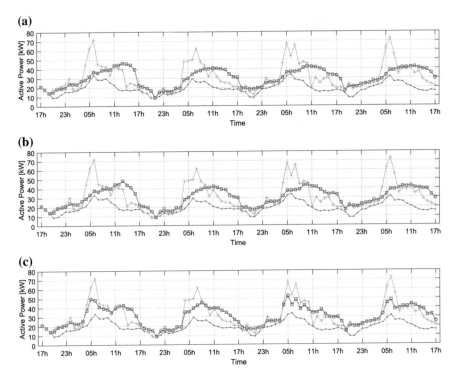

Fig. 4.21 **a** Comparison of: base load forecast (· marker), PEV aggregated load without management (+marker), and PEV aggregated load with MSD and full sets of MSs (□ marker). **b** Similar profiles using reduced sets of convenient MSs. **c** Similar profiles using reduced sets of the most convenient MSs, privileging early high charging rates. © [2017] IEEE. Reprinted, with permission, from [Ova+16b]

of charge are allowed to be between 30–80%. Among the states of charge profiles of Fig. 4.22 it is possible to notice bold profiles representing two PEVs in the three cases (of full and reduced sets of MSs), one with a 3.3 kW charger (in orange) and the other with a 7.5 kW charger (in blue). Comparing the bold profiles for cases with full sets and reduced sets of convenient MSs, it is possible to notice that using reduced sets (like those of Fig. 4.17) marginally reduces the capability of the proposed MSD approach for reducing the impact on the distribution system. The resulting profiles for both cases are similar, except for the fact that with reduced sets of MSs, PEVs are able to reach higher states of charge earlier in the connection period of time. On the other hand, for the same two PEVs when reduced sets of the most convenient MSs are employed (bottom chart in Fig. 4.22), the use of high charging rates is less constrained. As a result the states of charge reach higher values much earlier in the connection period of time (an advantage for vehicle owners). Besides, the MSD approach is still able to reduce the impact on the transformer, comparing with the case where the load of PEVs is not managed at all.

(a)

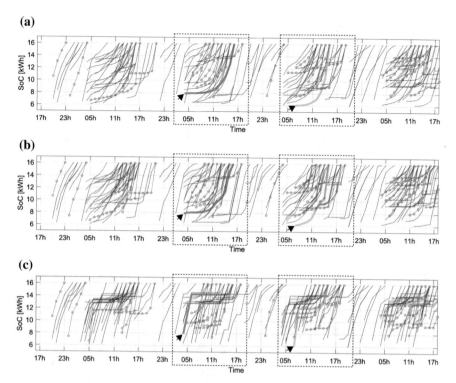

(b)

(c)

Fig. 4.22 **a** State of charge profiles (red without markers for 3.3 kW, and blue circle markers for 7.5 kW chargers) for case with full sets of MSs. **b** Similar profiles, for MSD with reduced sets of convenient MSs. **c** Similar profiles, for MSD with reduced sets of the most convenient MSs, privileging early high charging rates. © [2017] IEEE. Reprinted, with permission, from [Ova+16b]

Reduced sets of the most convenient MSs allow to take advantage from fast charging infrastructures while the impact on the transformers is still handled, as it is observed with the entropy maximization measurements on Fig. 4.23. Additionally, Fig. 4.24 shows the frequency distributions of time slots required to overpass 50% of the required energy, for all the PEVs during the four days. It can be observed that these frequency distributions are similar for full sets of MSs and sets of Best MSs, with mean values of 4.96 and 4.92 (time slots), respectively. However, with sets of Reduced Best MSs early fast charging rates are privileged, showing a reduction of the mean value to 3.28 time slots, and a concentration of more than 70 vehicles (over a total of 178) using only 2 time slots to overpass the 50%.

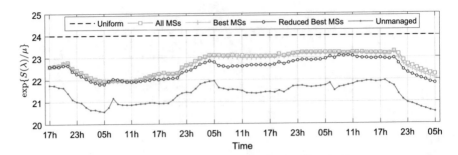

Fig. 4.23 Entropy measurements of total load distributions λ (Forecast + PEVs), for the unmmanaged case, the ideal case (uniform distribution), and for the MSD approach with full sets of MSs, and convenient MSs. Measurements are taken each hour on three of the four study days (with 24 hours horizons). © [2017] IEEE. Reprinted, with permission, from [Ova+16b]

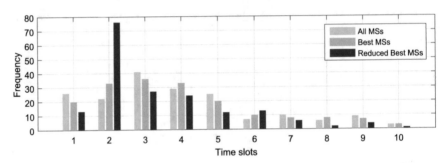

Fig. 4.24 Observed frequencies for the number of time slots required to overpass 50% of the energy needed. Frequencies corresponds to amounts of PEVs during the four study days. The MSD is tested using full sets of MSs, Best MSs, and Reduced Best MSs. © [2017] IEEE. Reprinted, with permission, from [Ova+16b]

4.6 Matlab Scripts

Most of the illustrative example results in this chapter were obtained using fixed step sizes in the discretization of the MSD equation. Using fixed step sizes requires forced routine stops when the step results in unfeasible distributions. However, for the application on PEV scheduling routines, a backtracking line-search subroutine, similar to that described in [BV04], is included to have adaptive step sizes and increase the routine's convergence speed and accuracy. In this case, the direction of descent employed in the backtracking line-search is provided by the MSD gradient. It is up to the reader to use this subroutine or not, and to test the differences in terms of performance.

This section provides Matlab scripts for the local MSD routines using full sets, and sets of the most convenient MSs. Sect. 4.6.1 describes a routine for computing the full set of MSs considering charger's nominal power, as well as a routine for obtaining subsets of the most convenient MSs. Sect. 4.6.2 describes the script for the local MSD routine. Finally, Sect. 4.6.3 provides an script for simulating the scenario, and the ideal operation of the aggregator, in the example of Sect. 4.5.2.

4.6.1 Scripts for Obtainig Sets of MSs

Let us recall the definitions of Sect. 4.3.2. The full set of MSs represents the vertices of the feasible region of local PEV load distributions that considers the charger's nominal charging rate. As it was mentioned, all the MSs are permutations of a vector of K^i elements whose first $(\gamma^i - 1)$ elements are all equal to $\overline{p}^i / \Gamma^i$; whose next element is equal to $(1 - \overline{p}^i (\gamma^i - 1)/\gamma^i)$; and whose final $(K^i - \gamma^i)$ elements are all equal to 0. It turns out that this vector is the single MS that provides the earliest and fastest charge for the PEV.

To obtain the full set of MSs from the *earliest—fastest* MS, the routine called perms_reps is employed. This routine was proposed in [D'E05] to recursively compute permutations of a vector with repeated elements. Notation and comments in the function have been adapted to fit the notation of the application proposed in this book.

```
function C = perms_reps(charging_rates,repetitions)
% Computes all the permutations of the fastest-earliest mixed strategy
%
% ------ Description inputs
% charging_rates : vector of charging rates in the fastest-earliest MSs
% repetitions : number of replicates of each rate in "charging_rates"
%
% ------ Description outputs
% C : Matrix of mixed strategies (full set)
%
% Example of inputs:
% if the fastest-earliest MS is [rate_1 rate_1 rate_2 0 0], then:
% - charging_rates = [rate_1 rate_2 0];
% - repetitions = [2 1 2];

% total number of elements
K_i = sum(repetitions);
n = length(charging_rates);

if K_i == 0
    C = [];
elseif K_i == 1
    C = charging_rates(repetitions>0);
else
    % there are at least two elements
    C = [];
    for i = 1:n
        if repetitions(i)>0
            repetitionst = repetitions;
```

```
        repetitionst(i) = repetitionst(i)-1;
        C_i = perms_reps(charging_rates,repetitionst);
        ni = size(C_i,1);
        C = [C;[repmat(charging_rates(i),ni,1),C_i]];
    end
  end
end
```

On the other hand, to compute only the set of best MSs from the *earliest—fastest* MS, the routine called find_best_MSs is employed. This routine computes a set of MSs using the same rules used to find those in Fig. 4.17. Parameter all_or_reduced determines whether the function returns a set with all the best MSs, or only a reduced subset of it (like those marked with a square in Fig. 4.17). It is important to mention that these rules for computing the set of best MSs are not general rules. Therefore, the reader is encouraged to update or improve the criteria for choosing the sets of the most convenient mixed strategies taking into account the desired application.

```
function C = find_best_MSs(charging_rates,repetitions,best_or_redbest)
% Computes all the permutations of the fastest-earliest mixed strategy
%
% inputs:
% charging_rates : vector of charging rates in the fastest-earliest MSs
% repetitions : number of replicates of each rate in "charging_rates"
% best_or_redbest : (1) all the best MSs are required
%                   (otherwise) only a reduced subset of the best MSs
%
% outputs:
% C : Matrix of mixed strategies (best set or reduced subset)
%
% Example of inputs:
% if the fastest-earliest MS is [rate_1 rate_1 rate_2 0 0], then:
% - charging_rates = [rate_1 rate_2 0];
% - repetitions = [2 1 2];

% total number of time slots
K_i = sum(repetitions);
slots = repetitions(1);

if K_i == 0
  C = [];
else
    % fastest-earliest MSs
    c_1 = [charging_rates(1)*ones(1,repetitions(1)) charging_rates(2)...
        zeros(1,repetitions(3))]';
    C = zeros(K_i,K_i-slots);
    C(:,1) = c_1;

    % creation of first block of the best MSs
    for mm = 2:K_i-slots
        C(:,mm) = circshift(c_1,[mm-1 0]);
    end
    if charging_rates(2) == 0
        C = [C circshift(c_1,[K_i-slots 0])];
    end

    % creation of second block of the best MSs
    if size(C,1) == size(C,2)
```

```
                 % C is the identity
        else
                 Cnext = C.*(ones(size(C,1),size(C,2))-eye(size(C,1),size(C,2)));
                 Cnext(1,:) = C(1,1)*ones(1,size(C,2));
                 if size(C,2) ~= 1
                          if best_or_redbest == 1
                                   C = [C Cnext(:,2:end)]; % All the best MSs
                          else
                                   C = [C(:,1:2) Cnext(:,2:end)]; % reduced subset
                                                     % of the best MSs
                          end
                 else
                          % C is equal to Cnext, then no additional MS is added
                 end
        end
end
C=C';
```

4.6.2 Script for the Local MSD Routine

The following is the local MSD routine that each PEV runs in order to find its optimal load distribution according to the parameters defined by the owner, and the load
forecast provided by the aggregator. As it was mentioned before, a backtracking line-
search subroutine, similar to that described in [BV04], is included to have adaptive
step sizes and increase the routine's convergence speed. In this case, the direction
of descent employed in the backtracking line-search is provided by the MSD gradient. It is up to the reader to use this subroutine or not, and test the differences in
terms of performance. To do this, it is possible to eliminate the code section named
backtracking line search, and replace it with a fixed value for the step
size Ts.

```
function [xhat, yhat] = Book_MSD_MEP(soc_d,soc_0,p_max,dt,Ki,Lhat_,...
                          x0,y0,alpha,reset_y0,...
                          all_best_or_redbest)

% ------ Desciption inputs
% soc_d : final desired state of charge [kWh]
% soc_0 : initial state of charge [kWh]
% p_max : max charging/discharging rate [kW]
% dt : time step
% Ki : total time steps
% Lhat_ : Load forecast from (t_0-1) to the avaiable horizon (>Ki+1)
% x0 : previously defined load distribution per pure strategies
% y0 : previously defined load distribution per mixed strategies
% alpha : owner's chosen trade-off factor in the interval [0,1]
% reset_y0 : (0) resets y0 to a uniform distribution
%               (otherwise) keeps the previously defined y0 as starting point
% all_best_or_redbest : (1) all the best MSs are required
%                       (2) only a reduced subset of the best MSs is required
%                       (otherwise) the full set of MSs is required

% ------ Description outputs
% xhat : optimal load distribution per pure strategies
```

```
% yhat : optimal load distribution per mixed strategies

% Gamma is the size of the "population" representing PEV load (sect.\,4.3.2)
Gamma = (soc_d-soc_0)/dt;

% slots+1 is the least amount of time slots required to consume the
% required amount of energy (soc_d-soc_0 )
slots = floor(Gamma/p_max); % gamma_i = slots+1 (sect.\,4.3.2)
if all_best_or_redbest == 1 || all_best_or_redbest == 2
    % best MSs or reduced subset
    C = find_best_MSs([p_max, Gamma - slots*p_max, 0]/Gamma,...
        [slots, 1, Ki-1-slots],all_best_or_redbest)';
else % all_best_or_redbest == 0 or any other value
    % All the MSs
    C = perms_reps([p_max, Gamma - slots*p_max, 0]/Gamma,...
        [slots, 1, Ki-1-slots])';
end
Ct = C';

% Number of mixed strategies
Mi = size(C,2);

% Number of internal MSD routine iterations
It = 1000;

% Definition of initial population distribution for Mixed Strategies
if isempty(y0)*(reset_y0~=0)+(reset_y0==0) == 1
    % this is the case if there is no previously defined distribution or
    % if it will be neglected
    msk = ones(1,Mi)*Gamma/Mi;
    mu = sum(Lhat_) + Gamma;
else
    % this is the case if a previous distribution has been defined, and it
    % will not be neglected
    msk = y0;
    mu = sum(Lhat_);
end

% the portion of the forecast that will interact with the PEV
Lhat = Lhat_(2:Ki+1);

% Initializes vector of population distribution for pure strategies
psk = (C*msk')';
% psk_ is a test variable before assignation to psk (useful inside routine)
psk_ = (C*msk')';

% Tolerance for the variance of payoffs
tol = 1e-10;
% Parameters of descent, and backtracking line search
alpha_ls = 0.5;
beta_ls = 0.1;

% Load forecast without local PEV load
bc = (Lhat - x0)';

% Payoff functions for pure strategies
fk = -1 - alpha*log(C*msk'/mu+bc/mu) - (1-alpha)*log(C*msk'/Gamma);

% Scalar field corresponding to payoff functions (based on sect.\,4.3.1)
F = -(alpha*(C*msk'+bc)'*log(C*msk'/mu+bc/mu)...
    -(1-alpha)*(C*msk')'*log(C*msk'/Gamma));
```

```
% Payoff functions for mixed strategies
gk = Ct*fk;

% weighted average payoff
fmeank = msk*gk/sum(msk);
if isempty(y0) == 0
    % if there is a previously defined distribution or it has already been
    % reset, the routine is run
    for i = 1:It

        % the new distribution is assigned
        psk = psk_;

        %% Subroutine for the backtracking line search
        Ts = 1e5;

        F_ir = Inf;
        % the descent direction (Delta_y) is defined by the MSD gradient
        Delta_y = msk.*(gk'-fmeank);
        if norm(Delta_y(:)) == 0
            % it only occurs when there is only one MS (i.e. gk' = fmeank)
            % or when the equilibrium has been perfectly reached
            break
        else
            Delta_y = Delta_y/norm(Delta_y(:)); % normalize the MSD gradient
        end

        while (F_ir > (F - alpha_ls*Ts*(fk'*(C*Delta_y'))))
            msk_ir = msk + Ts*Delta_y; % test distribution MSs
            psk_ir = (C*msk_ir')'; % test distribution for pure strategies

            % checks if test distributions lie within specified ranges
            if (sum(psk_ir<0) + sum(psk_ir>p_max) + sum(msk_ir<0) ...
                + sum(msk_ir>Gamma))>0
                % it is useless to update the scalar field value if
                % parameters are out of range
            else
                % updates scalar field test value
                F_ir = -(-alpha*(C*msk_ir'+bc)'*log(C*msk_ir'/mu+bc/mu)...
                    -(1-alpha)*(C*msk_ir')'*log(C*msk_ir'/Gamma));
            end
            % reduces step size Ts
            Ts = beta_ls*Ts;

            % If Ts is too small, the subroutine stops
            if Ts < 1e-30
                break
            end
        end
        if Ts < 1e- if Ts is too small, the MSD routine is stopped
            break
        end
        %%

        % Computes next distribution for Mixed Strategies
        msk = msk + Ts*Delta_y;

        % Computes next distribution for Pure Strategies
        psk_ = (C*msk')';

        % updates payoff functions for pure strategies
        fk = -1 -alpha*log(C*msk'/mu+bc/mu) -(1-alpha)*log(C*msk'/Gamma);
```

```
        % Scalar field corresponding to payoff functions (sect.\,4.3.1)
        F = -(-alpha*(C*msk'+bc)'*log(C*msk'/mu+bc/mu)...
            -(1-alpha)*(C*msk')'*log(C*msk'/Gamma));

        % updates payoff functions for mixed strategies
        gk = Ct*fk;

        % updates weighted average payoff
        fmeank = msk*gk/sum(msk);

        % Updates the weighted variance criteria
        variancef = sum(msk.*(gk'-sum(msk.*gk')/sum(msk)).^2)/sum(msk);

        % stopping criteria: variance, deviations, or max iterations
        if ( variancef < tol) || (((sum(msk)/Gamma)-1)^2>0.01)
            break
        end
    end
end
% output assignation
xhat = psk;
yhat = msk;
end
```

4.6.3 Script for the Simulation Scenario

The following is the data employed in the second example of Sect. 4.5.2. We encourage the reader to play with the inputs of the example to validate or improve the provided scripts.

```
close all
clear
clc

%%% Inputs

% base forecast load profile (it assumed to be fixed for the test days)
load_profile = [21.354 18.44 14.46 9.22 10.014 12.858 15.866 15.676 16.393 ...
                17.383 19.976 21.445 27.88 33.35 28.618 28.519 29.857 25.588 ...
                22.918 18.096 17.061 17.621 16.875 17.703 18.048 14.717 ...
                13.667 8.859 8.885 10.628 13.558 12.202 12.924 14.333 15.806 ...
                18.117 23.251 29.523 26.072 25.533 26.33 23.409 20.371 ...
                19.214 17.686 17.963 18.234 17.38 18.169 15.08 14.804 9.25 ...
                8.763 13.021 17.711 18.415 19.376 21.394 22.862 25.854 ...
                32.08 33.611 28.214 27.255 26.323 21.072 18.537 15.805 ...
                14.672 15.429 15.032 15.817 17.023 16.018 13.463 8.45 8.502 ...
                12.603 16.101 16.549 19.13 20.351 21.768 23.084 28.251 ...
                31.405 25.655 26.132 25.476 22.054 20.446 18.017 15.826 ...
                15.138 14.76 16.253 15.694 12.944 12.62 8.269 8.882 12.427 ...
                16.721 17.656 17.923 19.133 19.273 21.52 24.535 17.378 ...
                12.226 11.371 10.003 9.603 10.35 8.381 8.211 8.411 8.391 ...
                8.391 8.695 8.84 9.5 10.066 10.18 14.556 19.997 19.964 ...
                20.277 21.2 19.759 19.918 22.676 16.233 9.393 9.722 8.931 ...
                8.002 7.947 7.546 7.063 7.088 7.01 6.757 6.608 7.376 7.927 ...
                8.491 8.272 10.685 14.338 13.645 12.498 12.821 14.143 15.104 ...
```

```
                 19.813 24.689 21.712 23.606 25.006 22.617 20.151 17.385 ...
                 15.36 15.152 14.431 15.066];

% Time steps of arrival for each of the 323 PEVs in the considered days
t_arr = [4 4 5 5 7 8 8 8 11 12 12 12 12 12 12 12 13 13 13 13 14 14 14 14 14 ...
         14 14 16 16 16 17 18 19 19 19 19 19 20 20 20 20 21 21 22 23 24 ...
         30 30 31 34 36 36 36 36 36 36 36 36 37 38 38 38 38 39 39 39 ...
         39 39 39 40 40 41 41 42 43 43 43 44 44 45 46 47 51 51 52 54 56 ...
         56 57 60 60 60 60 60 60 61 61 61 61 61 63 63 63 63 63 63 64 64 ...
         64 65 65 65 65 66 66 68 68 68 69 69 70 70 71 71 71 71 74 76 76 ...
         76 76 79 80 82 84 84 84 84 85 85 85 85 85 85 85 86 86 86 86 86 ...
         86 86 87 87 87 87 88 88 89 89 89 90 90 90 91 91 92 92 92 92 93 ...
         93 93 94 94 95 96 96 97 98 99 104 104 108 108 108 108 108 108 ...
         108 108 108 108 109 109 109 109 110 110 110 110 111 112 112 112 ...
         112 112 112 113 114 114 115 115 115 115 116 116 116 117 118 118 ...
         118 118 119 119 120 120 123 125 129 129 130 130 130 131 132 132 ...
         132 132 133 133 133 133 133 133 133 134 134 134 134 134 135 135 ...
         135 135 135 135 135 136 136 137 137 137 138 138 138 139 139 139 ...
         140 140 140 141 142 143 147 148 150 150 151 151 153 154 155 156 ...
         156 156 156 156 156 156 156 156 156 157 157 157 157 157 157 158 ...
         158 158 158 158 158 158 159 159 159 159 160 161 161 162 163 163 ...
         164 164 166 166 166 168]';

% Time steps of departure for each of the 323 PEVs in the considered days
t_dep = [8 6 9 9 10 11 11 10 16 20 20 21 21 22 21 19 23 23 23 20 24 23 ...
         20 19 19 21 22 21 21 22 22 22 23 24 23 23 27 24 22 24 23 24 26 ...
         28 28 34 32 33 38 40 46 41 45 42 41 42 44 47 42 49 44 42 40 46 ...
         46 44 44 45 41 43 46 46 45 47 47 49 46 53 48 49 49 51 54 53 56 ...
         57 60 59 59 67 70 65 67 68 67 66 69 67 69 67 67 71 66 70 69 73 ...
         70 72 68 69 69 70 68 68 70 70 72 73 74 73 74 74 73 74 77 73 78 ...
         80 81 79 81 83 84 86 88 93 90 91 91 90 89 96 90 92 95 91 91 95 ...
         92 90 91 92 95 93 89 92 92 95 93 95 93 96 93 93 94 95 96 98 95 ...
         98 96 97 95 97 99 98 100 100 99 102 103 106 108 116 115 118 118 ...
         114 111 117 110 115 118 113 113 114 117 115 113 118 114 114 117 ...
         116 118 117 114 117 116 118 121 121 117 121 121 121 119 119 120 ...
         122 122 122 121 123 123 123 123 125 129 134 132 133 133 132 135 ...
         139 141 139 143 137 144 140 137 142 140 142 141 136 139 143 141 ...
         141 141 142 143 139 138 142 139 139 140 142 139 140 144 144 143 ...
         143 147 145 144 145 145 146 147 149 150 153 154 154 154 156 157 ...
         159 163 164 165 162 158 165 167 161 167 167 165 164 162 164 164 ...
         167 162 162 169 162 165 162 166 164 165 166 166 168 168 167 167 ...
         167 168 170 168 170 173 168 170]';

% Initial states of charge  for each of the 323 PEVs in the considered days
soc_0 = [0.9 4.6 2.3 0.3 6.2 1.4 5.5 6.9 0.5 6.9 3.7 1.9 5.6 3.3 0.6 2.7 ...
         2.5 0.1 0.7 6.8 3.2 7.3 0.3 6.8 4.9 1.9 0.2 3.6 2.3 6.3 5 4.8 ...
         5.3 1.4 5.2 3.3 3.4 5.2 7.1 5.8 2.7 2 1.1 5.1 4.2 0.7 7.7 7 1.9 ...
         4 3.6 7.8 2.5 4 0.4 6.9 6.2 1.9 3.7 0.5 0.5 3.1 5.3 7.1 0.6 7.1 ...
         2.9 2 3.7 4.1 2 3.3 1.2 1.2 7.8 1.6 1.4 1.8 7.6 0.3 7.9 0.8 3.8 ...
         6.9 1.4 5.6 2.9 1.7 7.4 0.2 1.8 5.1 3.8 2.1 3.1 1.1 0.5 3 5.4 ...
         0.8 6.3 6.2 6.7 0.1 1.6 3 2.6 2.6 2.3 6.5 3.3 4.2 4 2.8 6.3 ...
         2.9 3 3.6 3.6 7.1 1.2 7.2 6.8 1.5 7.6 1.3 7.3 0 4.1 0.1 0.8 4.4 ...
         1.2 4.7 0.3 5.2 3.1 5 4.3 5.7 3.3 7 1.2 4.4 3.3 1.6 3.4 7.2 7.4 ...
         1.9 4.3 8 6.7 6.8 3.7 0.8 6.2 6.3 6.8 1.2 1 7.5 5.3 3.3 2 3.7 ...
         3.9 3.5 0.2 6.9 2.6 8 5.3 5.4 3.3 4.7 3.2 4 1.2 2.2 7.5 2.2 6.6 ...
         6.8 5.2 0.2 4.3 4.5 6.1 7.8 4.6 6 7.9 5.7 2.1 6.5 5.5 5.4 1.4 ...
         7.9 2 1.2 4 0.6 2.5 5.5 0.8 6.3 1.5 2.2 2.5 4 1.1 1.2 0.7 4.2 ...
         4.5 4.4 2.7 1.7 2.4 6.5 5 5.9 7.2 5.2 6.8 5.8 3.7 4.2 4.2 4.4 ...
         7.2 2.1 2.9 5.1 4 1.1 4.4 0.8 1.4 0.4 4.5 6.5 5.3 2.6 6.4 2.7 ...
         7.9 0.8 6.5 3.6 7.6 3.2 0.8 3.8 0.1 7 6 4.8 5.7 7.7 7.4 5.3 1.3 ...
         1 2.3 0.6 5.7 0.7 0.8 5.4 4.3 1.5 7.8 7.5 5.1 1.9 6.9 5.7 5.4 ...
         4.7 2.2 7.9 1.5 6 0.2 7.1 6.4 0.4 5.7 1.8 2.2 2.5 2.2 0.2 0.8 ...
```

```
                3.9 1.4 4.8 3 2.2 5.1 5.1 2.4 0.2 4.9 2.8 3.8 0.3 6.4 0.5 2.1 ...
                4.9 0.6 7.6 1 0.5 1.4 4.3 4.8 7.8]';

% Desired states of charge for each of the 323 PEVs in the considered days
soc_d = soc_0*0 + 10; % all of them have 20kWh batteries (with available
                      % capacities between 30% and 80%, i.e., 10kWh)

% Nominal rates of charge for each of the 323 PEVs in the considered days
p_max = [3.3 7.5 3.3 3.3 3.3 7.5 3.3 3.3 3.3 3.3 3.3 3.3 3.3 3.3 3.3 3.3 ...
         3.3 3.3 7.5 3.3 3.3 3.3 3.3 3.3 3.3 3.3 3.3 3.3 3.3 3.3 3.3 3.3 ...
         3.3 3.3 7.5 3.3 3.3 3.3 7.5 3.3 7.5 7.5 3.3 7.5 3.3 3.3 3.3 3.3 ...
         3.3 3.3 3.3 3.3 3.3 3.3 3.3 3.3 3.3 3.3 7.5 3.3 3.3 3.3 7.5 3.3 ...
         3.3 3.3 7.5 3.3 7.5 7.5 3.3 7.5 3.3 3.3 3.3 3.3 7.5 3.3 3.3 3.3 ...
         7.5 3.3 3.3 3.3 3.3 3.3 3.3 3.3 3.3 3.3 3.3 3.3 3.3 3.3 3.3 3.3 ...
         3.3 3.3 7.5 3.3 7.5 3.3 3.3 7.5 7.5 3.3 3.3 3.3 3.3 3.3 3.3 3.3 ...
         3.3 7.5 7.5 7.5 3.3 7.5 3.3 3.3 3.3 3.3 3.3 7.5 3.3 3.3 3.3 3.3 ...
         3.3 3.3 3.3 3.3 3.3 3.3 3.3 3.3 7.5 3.3 3.3 7.5 3.3 3.3 3.3 3.3 ...
         3.3 3.3 3.3 3.3 3.3 3.3 7.5 3.3 7.5 3.3 3.3 7.5 3.3 3.3 3.3 7.5 ...
         3.3 3.3 7.5 7.5 7.5 7.5 3.3 3.3 3.3 3.3 7.5 3.3 3.3 3.3 3.3 7.5 ...
         3.3 3.3 7.5 3.3 3.3 3.3 3.3 3.3 7.5 3.3 7.5 3.3 7.5 3.3 3.3 3.3 ...
         3.3 7.5 3.3 3.3 3.3 3.3 7.5 3.3 3.3 7.5 3.3 7.5 3.3 3.3 7.5 3.3 ...
         3.3 3.3 3.3 3.3 7.5 3.3 3.3 3.3 3.3 3.3 7.5 3.3 3.3 3.3 3.3 3.3 ...
         3.3 3.3 3.3 3.3 3.3 3.3 3.3 3.3 7.5 7.5 3.3 3.3 3.3 3.3 3.3 3.3 ...
         3.3 3.3 3.3 3.3 3.3 3.3 7.5 3.3 7.5 7.5 3.3 7.5 3.3 3.3 3.3 3.3 ...
         7.5 3.3 3.3 3.3 3.3 3.3 3.3 3.3 3.3 3.3 3.3 3.3 3.3 3.3 3.3 3.3 ...
         3.3 7.5 3.3 7.5 7.5 3.3 3.3 3.3 3.3 3.3 3.3 3.3 3.3 3.3 3.3 3.3 ...
         3.3 3.3 3.3 3.3 3.3 3.3 3.3 3.3 3.3 3.3 3.3 3.3 7.5 3.3 3.3 3.3 ...
         3.3 3.3 3.3 3.3 3.3 3.3 3.3 3.3 3.3 3.3 3.3 3.3 3.3 3.3 3.3 3.3 ...
         3.3 7.5 3.3]';

% summary of arrival and departure times
t_temps = [t_arr t_dep];

alpha_avg = 0.95; % it is assumed that the trade-off factor is common
                  % to all PEVs. However the reader is encouraged to
                  % change this common value or randomize ir following
                  % any rule

% Total number of steps on the load profiles
T = length(load_profile);

% Time step duration (in hours or fractions)
dt = 1;

% Number of steps in a day
T_day = 24;

% Initial hour in the intial day (5pm in this case)
h_0 = 17; %hours (values from 0h to 23h)

% Total number of PEVs during the test days
count = length(p_max);

% Array where PEV load is individually placed by the aggregator
xhat = zeros(count,T);

% auxiliary variable for identifying the connected PEVs
index = 0;
% Array for storing the evolution of the initial state of charge which
% increases with the evolution of time
soc_0D = soc_0;
```

```
% Auxiliary variable representing the local memory of each PEV storing its
% current optimal distributionown
xy_Memory = cell(count,1);

% Number of exchanges of information between the aggregator and the and the
% connected PEVs per each time step
Num_exchanges = 10;

% Loop simulating the evolution of time and the interactions between
% Aggregator and PEVs

for t=1:T-3*T_day % only three days are considered
    % connected vehicles
    index = find(t>=t_temps(:,1) & t<=t_temps(:,2));
    if isempty(index)
        % no PEVs connected
    else
        for exchange = 1:Num_exchanges % Num_exchanges of info per time step
            for j = 1:length(index)
                if exchange == 1 % the start of a new round of exchanges
                    % PEVs erase their stored distributions per mixed
                    % strategies because each time a new round starts it
                    % means that the connection time left (Ki) has
                    % decreased and also the amount of MSs
                    xy_Memory{j,1} = [];
                end
                % the number ID associated to the current PEV
                m = index(j);
                % the aggregatr aggregates the current PEV load profiles to
                % the forecast and send it to the current PEV
                aggr_load_prof = load_profile+sum(xhat);

                % the aggregator verifies if the PEV is able to
                % redistribute its load or if it is already suing its time
                % slots left to the fullest (p_max)
                %
                % Variable (slots +1) is the amount of time slots required
                % to reach the desired state of charge
                slots = floor((soc_d(m)-soc_0D(m))/(dt*p_max(m)));
                if (t_dep(m)-t+1>=slots+1 && (t_dep(m)-t+1)>1)
                    % if the PEV is able to redistribute, then the
                    % aggregator sends the information. Then the PEV runs
                    % its local MSD routine :

                    % These input parameters of the local MSD routine
                    % can be changed

                    reset_y0 = 0; % keep previously stored distribution per MSs
                    all_best_or_redbest = 0; % use all the MSs

                    [xhat(m,t:t_dep(m)), xy_Memory{j,1}] = ...
                              Book_MSD_MEP(soc_d(m),...
                                soc_0D(m),p_max(m),dt,t_dep(m)-t+1,...
                                aggr_load_prof(t-1:T),xhat(m,t:t_dep(m)),...
                                xy_Memory{j,1},alpha_avg,reset_y0,...
                                all_best_or_redbest);
                else
                end
            end

            % plot base load forecast and aggregated PEV load
            plot(load_profile(:,t:t+T_day-1)',':','linewidth',1)
```

```
hold on
ProfilAux=load_profile'+sum(xhat)';
plot(ProfilAux(t:t+T_day-1,:),'linewidth',1)
hold off
grid
ylabel('Power [kW]')
xlabel('Forecast horizon [h]')
set(gca,'FontSize',12);
title( ['day = ' num2str(ceil((t*dt+h_0)/24)) '     hour = '...
   num2str(mod((t-1)*dt+h_0,24)) 'h    Exchange = ' ...
   num2str(exchange) '      # of PEVs = ' num2str(length(index))])
pause(1e-6)

end
% update the initial states of charge of the connected PEVs
% for the next time step
soc_0D(index)=soc_0D(index)+xhat(index,t);
end
end
```

This script has an embedded subroutine that shows the evolution of the load aggregation through time, given the interaction and information exchanges between aggregator and PEVs. We encourage the readers to plot variables like the final load distributions of each PEV, or their final state of charge trajectories. To do this, it is important to remember that the final load distributions for each PEV are stored in the array called xhat, and the state of charge trajectories can be obtained by doing a cumulative sum over xhat*dt.

4.7 Conclusion

An application of an Evolutionary Game dynamics called Mixed Strategist dynamics (MSD), for the decentralized load scheduling of Plug-in Electric Vehicles (PEV), is proposed. Following an analogy with the Maximum Entropy Principle (MEP) for tuning parameters of discrete probability distributions, entropy of the total load distribution and local PEV load distributions are considered as objectives of the scheduling approach. A trade-off among these entropy measurements is defined by PEV owners' convenience. While entropy maximization for the local load distributions contributes to preserve the batteries' states of health, entropy maximization for the total load distribution reduces undesirable load peaks on the transformer.

The problem is formulated such that final states of charge are assured depending on time constraints defined by owners. Furthermore, mixed strategies in the MSD are defined such that they represent the vertices of the convex set of feasible load distributions which results from constraints imposed by owners and chargers. The synergy of several PEVs is modeled as an application of the MSD in a multi-population scenario, where the interaction among populations follows another evolutionary game dynamics called Best Reply (BR) dynamics. The performance of the proposed approach is tested on real data measured on a distribution transformer from the SOREA utility grid company [SOR] in the region of Savoie, France.

This tool is specifically presented for the PEV load scheduling problem but can be applied to multiple problems of different nature. Multiple extensions of this type of methods can be developed specially in terms of the set of feasible solutions and the availability of information and communication infrastructure.

Furthermore, the selection of appropriate MSs in the MSD application for load management of PEV fleets is explored. The MSD approach using reduced sets of the most convenient MSs (privileging early high charging rates) is still able to handle the PEV load properly. Besides, fairness, PEV owners' convenience, and impact reduction are still achieved by the approach. As it has been studied, the most convenient MSs allow to take advantage from fast charging infrastructures while the impact on the transformers is still reduced, compared to the case without management.

In this distributed approach both the grid operator and PEV owners take advantage of the PEVs energy storage capacity. In this matter, trade-off factors and reduced the sets of MSs are both explored possibilities that provide important benefits to both actors of the problem. Both aspects of PEVs (as personal transportation systems, and as distributed energy storage devices) are exploited under the proposed distributed approach.

Chapter 5
Evolutionary Game Theory Approach
Part II: Escort Dynamics

In this chapter, the second part of an evolutionary game theory approach for the decentralized PEV load scheduling problem is presented. This approach is based on the application of a family of evolutionary game dynamics called Escort Dynamics (ED). In this application, a multi-population scenario is considered for representing PEV energy and reactive power quantities to be distributed over the three phases of the system and over multiple time slots in a given time horizon. The total number of populations is a function on the number of connected PEVs, while population sizes depend on owners' requirements. Depending on social or economic incentives from the utility grid manager, PEVs share the supply of several ancillary services to the grid: load shifting, active power balancing, and partial supply of reactive power demand per phase. Meanwhile, owners constraints are taken into account and batteries are guaranteed to be fully charged at the end of their charging time frame. In addition, chargers can be either three-phase or single-phase, and payoff functions are formulated such that resource and task allocation fairness is achieved for both types of chargers. Several numerical examples for single-phase, three-phase and both types of chargers are given. Performance is evaluated using real data from the SOREA utility grid company. The chapter ends with Matlab scripts for local ED routines handling energy populations and reactive power populations, for single and three-phase chargers. An additional script is provided for a simulation scenario used in multiple examples in the chapter.

5.1 Introduction

In Chap. 4, an evolutionary game dynamics approach was introduced based on the application of mixed strategist dynamics (MSD) in a multi-population approach. In general, this approach handles the distributed allocation of resources in a fair way for PEV owners, reducing the impact of charging PEVs to the grid. It takes advantage of the concept of mixed strategy and applies it to define local feasible regions as convex

© Springer International Publishing AG 2018
A. Ovalle et al., *Grid Optimal Integration of Electric Vehicles: Examples with Matlab Implementation*, Studies in Systems, Decision and Control 137, https://doi.org/10.1007/978-3-319-73177-3_5

hulls where each of the local[1] MSD dynamics evolve. However, this approach is still limited to the case of unidirectional power flow, where enumerating the whole sets of vertices (mixed strategies) is a computationally inexpensive procedure.

This chapter presents the theoretical base and details of an approach based on the application of a family of evolutionary game dynamics called Escort Dynamics (ED). The ED was defined in [Har11] based on information-geometric concepts. Its details, specially the concept of *escort function*, are explained in this chapter. After some interesting adaptations, the ED is able to represent local feasible regions as intersections of simplices. This results in multiple useful properties that motivate the application of this chapter, originally discussed in [Ova+17].

For instance, the multi-population application is now able to include both active and reactive power management, and handle both single and three-phase chargers. As with the MSD, in this approach all PEVs work together in a structure that fairly allocates resources and tasks. Additionally, PEVs are able to shift load in time, supply reactive power, and balance power quantities among phases of the transformer. PEV batteries are guaranteed to be fully recharged as well, according to the constraints imposed by its owners.

Finally, this chapter also provides Matlab scripts for the local ED routines employed by each PEV connected to either a single or a three-phase charger. The reader is encouraged to use and adapt these scripts to different scenarios, and propose alternatives to improve them.

5.2   Analogies Proposed in this Application

This chapter proposes a solution to a distributed resource allocation problem, dealing with several locally constrained variables. As in Chap. 4, the solution explained here is also a multi-population scenario. However, this time it is represented using the ED. As it will be observed, after some interesting adaptations, ED shows multiple useful properties that motivate this application. Some of the analogies employed in this chapter are similar to those introduced in Chap. 4. Nevertheless, given the flexibility achieved with ED, the multi-population model is now able to include both active and reactive power management, and handle both single and three-phase chargers. Taking this into account, let us introduce these analogies.

As in Chap. 4, here pure strategies also represent time slots. However, in this chapter they also represent the phases where quantities must be allocated. Populations, on the other hand, represent quantities (energy and reactive power) to be distributed among three phases in several time slots. The total number of populations depends on the number of connected PEVs, while the size of each population and the number of pure strategies for each population depend on the owners' requirements and the number of phases of the charger.

---

[1]The word *local* refers to each PEV individually.

Each PEV is locally in charge of two populations representing energy, and reactive power. Depending on the duration of connection, and the charger (single or three-phase), the dimension of simplices and feasible subsets for each PEV are defined.

5.2.1 Populations Representing Energy

Let us check Fig. 5.1a. In this diagram, three populations with a given number of individuals are distributed among 12 *territories* identified by pairs (k, m). Each territory has a given *hosting capacity* for individuals of the populations, and each territory provides a certain payoff for those individuals. Among these three populations, the first one has a *sedentary* behavior, i.e., its individuals are not willing to migrate to other territories. The other two populations have a *nomad* behavior, i.e., their individuals are prone to move among territories to increase their population's payoff. Each territory represents a pure strategy for individuals of nomad populations.

The total number of individuals of the three populations does not change with time, i.e. the size of the total population is constant through the time. However, the nomad populations are able to *recruit* individuals from the sedentary population. Thus, the size of the nomad populations may increase with time at the expense of the sedentary population size decreasing. Nonetheless, the size of the two nomad populations is limited by given *reception capacities*.

Figure 5.1a shows initial sizes and distributions of the three populations on the left. After evolution (on the right) these sizes and distributions change according to the payoffs provided by the 12 territories until an evolutionary stable state[2] is reached. It may be noticed, for instance, that after evolution territories like $(k = 2, m = 2)$ lose most of their sedentary population quotas. Also, the total population composition passes from an initial domination of sedentary population, to an almost even composition after evolution, slightly dominated by one of the nomad populations.

This scenario summarizes the analogies followed in the approach of this chapter. Each territory represents a phase $m = \{1, 2, 3\}$ at time steps $k = \{1, 2, 3, 4\}$. The sedentary population represents the forecasted energy consumption per phase of the transformer and time slot. Hosting capacities of territories represent the load limits of the transformer at each phase. Nomad populations represent the energy required by each PEV, and the size of the population represents the amount of energy required to reach a desired state of charge. The recruitment idea represents the possibility of connected PEVs consuming energy from one phase at a given instant, and re-injecting it at another phase and another instant. The reception capacities of a nomad populations represent PEV storage limits. Nominal charging rate limits for each PEV charger are represented by the limit amount of individuals that a nomad population is able to allocate in or recruit from a territory. Finally, the idea behind the evolution of the distribution of nomad populations reflects the fact that PEVs may allocate energy

---

[2]The concept of evolutionary stable state (ESS) is described later in this chapter.

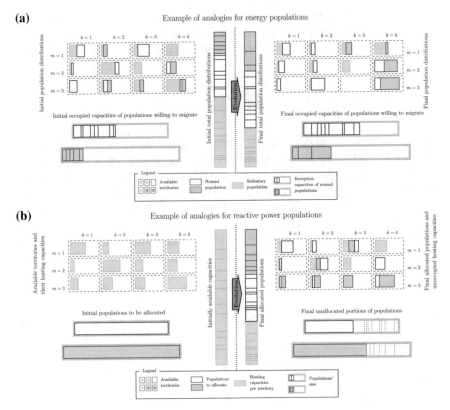

Fig. 5.1 Illustrative example of the multi-population approach proposed for the distributed energy management of electric vehicles

consumption/injection among phases and time slots depending on benefits they can get from, or services they can provide to the grid.

5.2.2 Populations Representing Reactive Power

Now, let us consider the diagrams on Fig. 5.1b. In this case, initially a certain *hosting capacity* is predefined for each of 12 *territories* identified by pairs (k, m), and again each of them provides a certain payoff. On the other hand, let us consider two populations wanting to allocate their individuals among the initially unoccupied territories. Populations want to allocate their individuals maximizing their payoff but taking into account the available hosting capacities of each territory. Each population has a predefined amount of individuals it may choose to allocate or not on each territory, depending on the payoff they get. Given this conditions, the amount of allocated individuals by each population, per territory, evolves in time until an evolutionary stable state is reached. Again, each territory represents a pure strategy for each population.

Figure 5.1b shows initial available hosting capacities, and final allocated portions of the two populations according to the payoffs provided by the 12 territories. When the equilibrium is reached, almost half of the first population and a third of the second are allocated in the territories. Besides, almost two thirds of the total hosting capacity is taken by the populations. The remaining one third is not taken because an equilibrium has been reached. This means that if populations allocate more individuals, their payoff will decrease.

This scenario summarizes the analogies proposed for the reactive power management application of this chapter. Each territory represents a phase $m = \{1, 2, 3\}$ at time $k = \{1, 2, 3, 4\}$. The hosting capacities of each territory represent the forecasted reactive power per phase of the transformer and time slot. A population represents the portion of the forecasted reactive power demand that a PEV is able to supply. The size of a population represents the amount of reactive power that the corresponding PEV is able to supply, given its energy consumption and its nominal rate limits (charger constraints). Finally, the idea of allocating individuals of each population to the available territories according to their hosting capacities, represents PEVs supplying the reactive power demand of the grid.

The next section describes the ED and its similarities with the MSD. This is followed by the description of the proposed *intersection escort functions* that allow to represent the analogies above using the ED.

5.3 Introduction to the Escort Evolutionary Game Dynamics

The ED describes the evolution of the distribution of a normalized population over K possible pure strategies, according to the benefit provided by those strategies. This generalized dynamics is defined in continuous time by

$$\dot{x}_k = \phi_k(x_k)(f_k(\mathbf{x}) - \bar{f}_\phi(\mathbf{x})), \tag{5.1}$$

where $\mathbf{x} = [x_1, x_2, \ldots, x_k, \ldots, x_K]^{\mathrm{T}}$ is the state vector of portions of the population following pure strategy k, $f_k(\mathbf{x})$ is the payoff function[3] for strategy k, and $\bar{f}_\phi(\mathbf{x})$ is the weighted average payoff. The so called *escort function* $\phi_k(x_k)$ can be understood as an incentive for the rate of growing.

From (5.1) it can be noticed that: if the escort $\phi_k(x_k)$ is initially zero, x_k will remain unchanged for $t > 0$; portions with payoff functions greater that the weighted average payoff tend to grow and vice-versa; an equilibrium is reached when $f_k(\mathbf{x}) = \bar{f}_\phi(\mathbf{x})$. The weighted average payoff can be expressed as

---

[3]Elements f_k are completely different and should not be confused with the nomenclature employed on the dynamic programming algorithm of Chap. 3, for the model of the system.

$$\bar{f}_\phi(\mathbf{x}) = \frac{1}{\Phi(\mathbf{x})} \sum_{k=1}^{K} \phi_k(x_k) f_k(\mathbf{x}), \tag{5.2}$$

where $\Phi(\mathbf{x}) = \sum_{k=1}^{K} \phi_k(x_k)$. If positive escort functions are considered, (5.2) can be understood as the expected value of the payoff functions given a probability distribution defined by the escort functions as,

$$\check{\phi}(\mathbf{x}) = \frac{1}{\Phi(\mathbf{x})} [\phi_1(x_1), \phi_2(x_2), \dots, \phi_K(x_K)]^{\mathsf{T}}.$$

Summing (5.1) over all the portions of the population provides

$$\sum_{k=1}^{K} \dot{x}_k = \sum_{k=1}^{K} \phi_k(x_k) f_k(\mathbf{x}) - \bar{f}_\phi(\mathbf{x}) \sum_{k=1}^{K} \phi_k(x_k)$$

$$= \Phi(\mathbf{x}) \bar{f}_\phi(\mathbf{x}) - \Phi(\mathbf{x}) \bar{f}_\phi(\mathbf{x}) = 0. \tag{5.3}$$

Therefore, if at $t = 0$ the sum of portions of population $\sum_{k=1}^{K} x_k = 1$, then the sum of portions will remain equal for $t > 0$. Since ED assures $\sum_{k=1}^{K} x_k = 1$ for $t > 0$, the definition of the escort functions $\phi_k(x_k)$ is crucial if it is desirable to keep the state vector $\mathbf{x}$ within a more restrictive region, subset of the plane $\sum_{k=1}^{K} x_k = 1$ in $\mathbb{R}^K$. For instance, if escort functions are $\phi_k(0) = 0$ and positive for $x_k > 0$, the inequality $x_k \geq 0$ becomes a *barrier* for (5.1). Thus, if $x_k > 0$ for all k at $t = 0$, the states $\mathbf{x}$ at $t > 0$ given by the ED, will be within the convex set Δ^K, usually known as the standard simplex, defined as

$$\Delta^K = \left\{ \mathbf{x} \in \mathbb{R}^K : x_k \geq 0, \sum_{k=1}^{K} x_k = 1 \right\}.$$

Such is the case for $\phi_k(x_k) = x_k$, where the incentive is proportional to the portion of the population. In this case, pure strategies are represented by the canonical base of $\mathbb{R}^K$ which are the vertices of the simplex. For increasing escort functions such that $\phi_k(0) = 0$, the standard simplex is invariant under the ED.

5.3.1   ED as a Gradient Flow

For the simplex scenario, the author of [Har11] establishes several useful definitions. The standard simplex is a $(K - 1)$-dimensional Riemmanian manifold embedded in $\mathbb{R}^K$. Its tangent space is defined as $T_\mathbf{x}\Delta^K = \{\mathbf{w} \in \mathbb{R}^K : \sum_{k=1}^{K} w_k = 0\}$ [Leb06]. Equipped with the so called *Escort metric*, the local *Escort inner product* of two vectors $\mathbf{a}$ and $\mathbf{b}$ in the tangent space $T_\mathbf{x}\Delta^K$ is defined by $< \mathbf{a}, \mathbf{b} >_\phi = \sum_{k=1}^{K} a_k b_k / \phi_k(x_k)$.

It differs from the euclidean inner product in the sense that it depends on the position of $\mathbf{x}$ and the corresponding value of the escort functions.

As it is for the definition of the euclidean gradient $\nabla F(\mathbf{x})$ with the euclidean inner product, the *Escort gradient* vector $\nabla_\phi F(\mathbf{x})$ is uniquely defined as the vector whose escort inner product with a given vector $\mathbf{w} \in T_\mathbf{x}\Delta^K$ produces the directional derivative $D_\mathbf{x} F(\mathbf{w}) = < \nabla_\phi F(\mathbf{x}), \mathbf{w} >_\phi$ of the potential function $F(\mathbf{x}) : \Delta \to \mathbb{R}$, along $\mathbf{w}$.

Given (5.3), the dynamics (5.1) represents the elements of a vector in the tangent space $T_\mathbf{x}\Delta^K$. Let us call this vector θ. As it was defined before, the escort inner product between θ and another vector $\mathbf{w}$ in the tangent space is,

$$< \theta, \mathbf{w} >_\phi = \sum_{k=1}^K \frac{\theta_k w_k}{\phi_k(x_k)} = \sum_{k=1}^K (f_k(\mathbf{x}) - \bar{f}_\phi(\mathbf{x})) w_k$$

$$= \sum_{k=1}^K f_k(\mathbf{x}) w_k - \bar{f}_\phi(\mathbf{x}) \sum_{k=1}^K w_k = \mathbf{f}(\mathbf{x}) \cdot \mathbf{w},$$

where $\mathbf{f}(\mathbf{x}) = [f_1(\mathbf{x}), f_2(\mathbf{x}), \ldots, f_k(\mathbf{x}), \ldots, f_K(\mathbf{x})]^T$ is the vector of payoff functions. Thus, if $\mathbf{f}(\mathbf{x})$ is the euclidean gradient of a potential function $F(\mathbf{x})$, i.e. $\mathbf{f}(\mathbf{x}) = \nabla F(\mathbf{x})$, then,

$$< \theta, \mathbf{w} >_\phi = \mathbf{f}(\mathbf{x}) \cdot \mathbf{w} = \nabla F(\mathbf{x}) \cdot \mathbf{w} = D_\mathbf{x} F(\mathbf{w}). \qquad (5.4)$$

Therefore θ, defined by (5.1), is an escort gradient vector $\nabla_\phi F(\mathbf{x})$. Let us define a local state $\hat{\mathbf{x}}$. For states $\mathbf{x}$ in some neighborhood of $\hat{\mathbf{x}}$, vectors $\hat{\mathbf{x}} - \mathbf{x}$ belong to the tangent space and point to $\hat{\mathbf{x}}$. It should be noticed that, if $\hat{\mathbf{x}}$ is a local maximum, then the directional derivative of the potential function $F(\mathbf{x})$ at a point $\mathbf{x}$, along $\hat{\mathbf{x}} - \mathbf{x}$, is always larger than zero, if $\mathbf{x} \neq \hat{\mathbf{x}}$. This can be expressed as,

$$D_\mathbf{x} F(\hat{\mathbf{x}} - \mathbf{x}) = (\hat{\mathbf{x}} - \mathbf{x}) \cdot \nabla F(\mathbf{x}) = (\hat{\mathbf{x}} - \mathbf{x}) \cdot \mathbf{f}(\mathbf{x}) > 0. \qquad (5.5)$$

In words, this means that if $\hat{\mathbf{x}}$ is an equilibrium point, the angle between the vector $(\hat{\mathbf{x}} - \mathbf{x})$ and the gradient vector $(\nabla F(\mathbf{x}) = \mathbf{f}(\mathbf{x}))$ is always an acute angle. Furthermore, it implies that the directional derivative at a point $\mathbf{x}$ in the direction of a vector pointing to the local optimum $\hat{\mathbf{x}}$ is always positive, except if $\mathbf{x} = \hat{\mathbf{x}}$. If $\mathbf{w}$ in (5.4) is replaced by $\hat{\mathbf{x}} - \mathbf{x}$, then

$$< \nabla_\phi F(\mathbf{x}), \hat{\mathbf{x}} - \mathbf{x} >_\phi = D_\mathbf{x} F(\hat{\mathbf{x}} - \mathbf{x}) = (\hat{\mathbf{x}} - \mathbf{x}) \cdot \mathbf{f}(\mathbf{x}) > 0,$$

which means that for the escort gradient defined with the escort inner product, the important inequality of (5.5) holds as well. Consequently, the ED is a gradient flow meaning that the trajectories from any initial state $\mathbf{x}(0)$ are such that the maximal

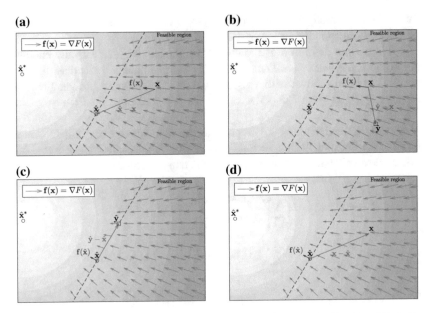

Fig. 5.2 **a** Example of an Evolutionary Stable State (ESS) $\hat{\mathbf{x}}$ as described by Eq. (5.5), inside a feasible region. The global optimum $\hat{\mathbf{x}}^*$ is outside the feasible region. The vector field inside the feasible region corresponds to the euclidean gradient (Feasible and unfeasible optima are marked with ◦). **b** Example of a random point $\hat{\mathbf{y}}$ (marked with □) inside the feasible region which does not fulfill the condition of an ESS. **c** Example of another random point $\hat{\mathbf{y}}$ (marked with □) in the boundary of the feasible region as well as the real ESS. It is closer to fulfill the ESS condition but it is still not the case because of the gradient evaluated on the real ESS. **d** Illustration (for comparison) of the definition of Nash Equilibrium for the same points of the illustration of ESS

direction of change (given by the escort gradient) is followed, and the local optimum $\hat{\mathbf{x}}$ is reached [HS88, HS98, Men+95].

The equilibrium state $\hat{\mathbf{x}}$ defined by the inequality (5.5) is also known as an Evolutionary Stable State (ESS). This concept can be illustrated with Fig. 5.2. In this example, a local maximum of a potential function $F(\mathbf{x})$, labeled as $\hat{\mathbf{x}}$, is at the boundary of the feasible region defined by some constraints. Given that the true optimal $\hat{\mathbf{x}}^*$ is outside the feasible region, the euclidean gradient at the local feasible optimum $\hat{\mathbf{x}}$ is $\nabla F(\hat{\mathbf{x}}) \neq \mathbf{0}$. It should be mentioned that the feasible optimal is the point where the boundaries of the feasible region touch the highest valued level set (level curve or level surface in lower dimensions) of the potential function. The ESS, as defined by Eq. (5.5), is illustrated in Fig. 5.2a, for a random point $\mathbf{x}$ in the neighborhood of $\hat{\mathbf{x}}$. For any point $\mathbf{x}$ in the neighborhood of $\hat{\mathbf{x}}$, the vector $\hat{\mathbf{x}} - \mathbf{x}$ forms an acute angle with the euclidean gradient vector. Thus, the dot product of these two vectors is always positive unless $\hat{\mathbf{x}} = \mathbf{x}$.

Condition (5.5) does not hold if $\hat{\mathbf{x}}$ is not the local optimum, as it is illustrated in Fig. 5.2b, for another randomly chosen sub-optimal point $\hat{\mathbf{y}}$ in the feasible region. The condition does not hold even for other points in the boundaries, different than the local optimum, as it is illustrated in Fig. 5.2c. Here, the condition does not hold because $\hat{\mathbf{y}} - \hat{\mathbf{x}}$ always forms a right angle with the euclidean gradient evaluated on the true local optimum $\hat{\mathbf{x}}$, given that the gradient is always orthogonal to the level sets. Thus, $(\hat{\mathbf{y}} - \hat{\mathbf{x}}) \cdot \mathbf{f}(\mathbf{x}) = 0$.

It is also important to clarify the link between the definition of ESS and Nash Equilibrium (NE). NE is commonly defined as,

$$(\mathbf{x} - \hat{\mathbf{x}}) \cdot \mathbf{f}(\hat{\mathbf{x}}) \leq 0. \qquad (5.6)$$

which relates the euclidean gradient evaluated at the equilibrium $\hat{\mathbf{x}}$ an a vector $(\mathbf{x} - \hat{\mathbf{x}})$ pointing to states $\mathbf{x}$ in the neighborhood of the equilibrium, as it is shown on Fig. 5.2d. These two vectors always form obtuse or right angles, resulting in zero or negative dot products. The ESS and the NE differ on the point of evaluation of the euclidean gradient. However, the ESS is a more refined definition of stable equilibrium state than the NE defintion, since the gradient evaluated at saddle points (unstable equilibria) is a zero valued vector satisfying the condition (5.6). As it can be observed, the ESS definition implies Strict NE (strict inequality) and vice-versa [HS98].

For non-decreasing escort functions, and strictly positive on $0 < x_k < 1$, the author of [Har11] provides Lyapunov functions for the ED, proving the stability of the ESS, which is a local maximum of $F(\mathbf{x})$. These Lyapunov functions are dependent on the escort functions definition [Har11, Har09, HS98, HS88].

5.4 Intersection Escort Functions

Escort functions proposed in [Har11] are strictly positive and non-decreasing on $0 < x_k < 1$ to properly define a local Lyapunov function for an equilibrium $\hat{\mathbf{x}}$ and the escort dynamic, in the standard simplex. In this chapter, the characteristics of escort functions are exploited by giving them alternative definitions to limit $\mathbf{x}$ to be in a more general set Ψ^K.

These new definitions are motivated by the development of a ED application in optimal distributed resource allocation scenarios like the objective in this book. The proposed escort functions are neither strictly positive nor non-decreasing on the interval $0 < x_k < 1$. However, a Lyapunov function can still be found for an equilibrium $\hat{\mathbf{x}}$ in the region Ψ^K. The first part of this section summarizes some information about the desirable feasible region. The second part provides the definition of the proposed *intersection escort functions* and the reason that motivates those definitions.

5.4.1 The Feasible Region

The kind of problems for the proposed application motivate a desirable feasible region Ψ^K of the form,

$$\Psi^K = \left\{ \mathbf{x} \in \mathbb{R}^K : x_k^{lo} \leq x_k \leq x_k^{up}, \sum_{k=1}^{K} x_k = 1 \right\},$$

where $x_k^{lo} < x_k^{up}$, and both can be larger or smaller than the unity for all k. This set can be defined as the intersection $\Psi^K = \Delta_{lo}^K \cap \Delta_{up}^K$, where Δ_{lo}^K and Δ_{up}^K are two simplices that lie in the same hyperplane $\sum_{k=1}^{K} x_k = 1$ as the standard simplex Δ^K. Δ_{lo}^K and Δ_{up}^K can be subsets of Δ^K (i.e. the standard simplex) depending on the corresponding limits x_k^{lo} and x_k^{up}. These simplices are defined as follows,

$$\Delta_{lo}^K = \left\{ \mathbf{x} \in \mathbb{R}^K : x_k \geq x_k^{lo}, \sum_{k=1}^{K} x_k = 1 \right\},$$

$$\Delta_{up}^K = \left\{ \mathbf{x} \in \mathbb{R}^K : x_k \leq x_k^{up}, \sum_{k=1}^{K} x_k = 1 \right\}.$$

The set Ψ^K, the simplices Δ_{lo}^K and Δ_{up}^K, and the standard simplex Δ^K, are all convex polytopes that can be represented by their set of vertices. As it can be noticed, all those vertices are points in the hyperplane $\sum_{k=1}^{K} x_k = 1$. For instance, the set of vertices for the standard simplex is the K-dimensional identity matrix $\mathbf{I}$ (i.e. the canonical base of $\mathbb{R}^K$). For the simplices Δ_{lo}^K and Δ_{up}^K the sets of vertices can be represented as the column vectors of matrices $\mathbf{C}_{lo}$ and $\mathbf{C}_{up}$ respectively. These matrices are K-dimensional, and represent bases for $\mathbb{R}^K$ as well.

On the other hand, the number of vertices for the set Ψ^K can be much larger that K and enumerating all of them is a combinatorial problem, hard to solve computationally [AF96]. For instance, on a simplified scenario where the limits are homogeneous, i.e. $x_k^{lo} = 0$ and $x_k^{up} = x^{up}$ for all k, as in [Ova+16a], the unity can be divided in γ pure strategies at most, where γ is an integer that depends on x^{up} and can be at least 1 and at most K. In this case, the amount of vertices for the subset can be computed by $\frac{K!}{(\gamma-1)!(K-\gamma)!}$. This amount is reduced for small values of γ close to 1 or large values close to K. However, even in this case, the number of vertices becomes extremely large for γ with values near $K/2$. Thus, for the general case, trying to find the vertices of Ψ^K is not a viable procedure.

Instead, matrices $\mathbf{C}_{lo}$ and $\mathbf{C}_{up}$ can be easily found. Let $\mathbf{x}^{lo} = [x_1^{lo}, x_2^{lo}, \ldots, x_K^{lo}]^T$ and $\mathbf{x}^{up} = [x_1^{up}, x_2^{up}, \ldots, x_K^{up}]^T$ be column vectors containing the constrain values of $\mathbf{x} \geq \mathbf{x}^{lo}$ and $\mathbf{x} \leq \mathbf{x}^{up}$, and defining the simplices Δ_{lo}^K and Δ_{up}^K respectively. Matrices $\mathbf{C}_{lo}$ and $\mathbf{C}_{up}$ are defined by

$$\mathbf{C}_{lo} = \mathbf{X}_{lo} + \sigma_{lo}\mathbf{I}, \quad \mathbf{C}_{up} = \mathbf{X}_{up} + \sigma_{up}\mathbf{I}, \tag{5.7}$$

where $\mathbf{X}_{lo}$ and $\mathbf{X}_{up}$ are square matrices with K copies of $\mathbf{x}^{lo}$ and $\mathbf{x}^{up}$ as columns, respectively. Scalars σ_{lo}, and σ_{up} are

$$\sigma_{lo} = 1 - \sum_{k=1}^{K} x_k^{lo}, \quad \sigma_{up} = 1 - \sum_{k=1}^{K} x_k^{up}.$$

The importance of these matrices lies in the fact that they can be employed to redefine equivalently these sets as convex hulls as follows,

$$\Delta_{lo}^{K} = \left\{ \mathbf{x} = \mathbf{C}_{lo}\beta : \beta \in \mathbb{R}^K, \beta_k \geq 0, \sum_{k=1}^{K} \beta_k = 1 \right\},$$

$$\Delta_{up}^{K} = \left\{ \mathbf{x} = \mathbf{C}_{up}\alpha : \alpha \in \mathbb{R}^K, \alpha_k \geq 0, \sum_{k=1}^{K} \alpha_k = 1 \right\}.$$

Figure 5.3 is useful to illustrate these convex sets and their intersection in $\mathbb{R}^3$. The first diagram on the top left illustrates the standard simplex Δ^3 and the simplex of lower constraints Δ_{lo}^3. The column vectors of $\mathbf{C}_{lo}$ are the vertices of the simplex Δ_{lo}^3. The diagram on the top right corresponds to the the standard simplex Δ^3 and the simplex of upper constraints Δ_{up}^3. This last simplex looks like it is *inverted*, this is given by the nature of its inequalities. In this case, the column vectors of $\mathbf{C}_{up}$ are the vertices of the simplex Δ_{up}^3 as well. The diagram on the bottom illustrates the intersection $\Psi^3 = \Delta_{lo}^3 \cap \Delta_{up}^3$. This diagram is useful to put some light on the fact that the amount of vertices of the intersection depends on the constraints and is not a simple problem to solve computationally.

For a known state $\mathbf{x} \in \Delta^K$, the corresponding representation on the sets Δ_{lo}^K and Δ_{up}^K in terms of the corresponding states β and α, can be computed based on the inverse matrices of $\mathbf{C}_{lo}$ and $\mathbf{C}_{up}$ respectively. From (5.7),

$$\beta = \mathbf{C}_{lo}^{-1}\mathbf{x} = \frac{1}{\sigma_{lo}}(\mathbf{I} - \mathbf{X}_{lo})\mathbf{x},$$

$$\alpha = \mathbf{C}_{up}^{-1}\mathbf{x} = \frac{1}{\sigma_{up}}(\mathbf{I} - \mathbf{X}_{up})\mathbf{x}.$$

Knowing that $\sum_{k=1}^{K} x_k = 1$, the expressions before can be simplified. The resulting expressions for vectors β and α from the known vector state $\mathbf{x} \in \Delta^K$ are

$$\beta = \frac{1}{\sigma_{lo}}(\mathbf{x} - \mathbf{x}_{lo}), \tag{5.8}$$

$$\alpha = \frac{1}{\sigma_{up}}(\mathbf{x} - \mathbf{x}_{up}). \tag{5.9}$$

Expressions (5.8) and (5.9) provide direct useful information of $\mathbf{x}$ approaching to the boundaries of $\Psi^K = \Delta_{lo}^K \cap \Delta_{up}^K$. It is important to notice that β_k and α_k depend only on x_k. Besides, vertices $\mathbf{C}_{lo}$ and $\mathbf{C}_{up}$ can be interpreted as the pure strategies for

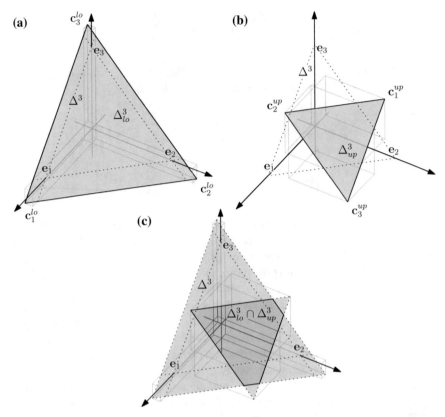

Fig. 5.3 Simplices in $\mathbb{R}^3$: **a** original simplex Δ^3, and simplex of lower constraints Δ_{lo}^3; **b** original simplex Δ^3, and simplex of upper constraints Δ_{up}^3; **c** Intersection of both simplices $\Delta_{lo}^3 \cap \Delta_{up}^3$. Example simplices for $K = 3$, and some random constraints

simplices Δ_{lo}^K and Δ_{up}^K respectively. Thus, the population of the k-th pure strategy in Δ_{lo}^K or in Δ_{up}^K depends only on the population of the k-th pure strategy in Δ^K.

Expressions (5.8) and (5.9) can be candidates for escort functions. Figure 5.4 shows some typical plots for $\phi_k(x_k) = \beta_k = (x_k - x_k^{lo})/\sigma_{lo}$ and $\phi_k(x_k) = \alpha_k = (x_k - x_k^{up})/\sigma_{up}$, the identity function $\phi_k(x_k) = x_k$ (replicator case), and an alternative plot $\phi_k(x_k) = \alpha_k \beta_k$ as well. It is important to notice that β_k and α_k reach the unity at $x_k = \sigma_{lo} + x_k^{lo}$ and $x_k = \sigma_{up} + x_k^{up}$ respectively, however β_k is monotonically increasing and α_k is monotonically decreasing. This occurs because $\sigma_{lo} > 0$ and $\sigma_{up} < 0$ are requirements always. Otherwise, constraints x_k^{lo} are too high or constraints x_k^{up} are too low and the unity (i.e. the total normalized population) is unfeasible. The point where β_k or α_k reach the unity, also corresponds to the point where the vertex of the corresponding simplex is placed.

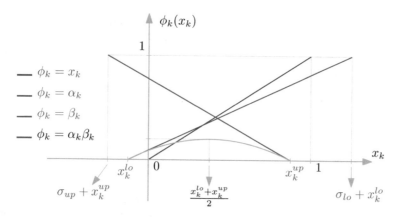

Fig. 5.4 Typical plots for the proposed escort functions ($\phi_k(x_k) = \alpha_k$, $\phi_k(x_k) = \beta_k$, and $\phi_k(x_k) = \alpha_k\beta_k$), and for the identity escort function ($\phi_k(x_k) = x_k$). ©[2017] IEEE. Reprinted, with permission, from [Ova+17]

5.4.2 Definition of the Proposed Escort Functions

As it was explained in Sect. 5.3, ED guarantees $\sum_{k=1}^{K} x_k = 1$. Given a continuous escort function $\phi_k(x_k)$, if at $t = 0$, x_k is such that $\phi_k(x_k) > 0$ for all k, then the escort dynamic will provide states **x** for $t > 0$ such that $\phi_k(x_k) \geq 0$ for all k. From (5.1), if $\phi_k(x_k)$ approaches to zero then $\dot{x}_k$ approaches to zero as well. Thus the zeros of the escort functions represent *barriers* for the feasible states **x**.

If $\phi_k(x_k) = \alpha_k$, and at $t = 0$, x_k is such that $\alpha_k > 0$ for all k, then the escort dynamic will provide states **x** such that $\alpha_k \geq 0$. Since it is guaranteed that $\sum_{k=1}^{K} x_k = 1$, from (5.9) it is also guaranteed that $\sum_{k=1}^{K} \alpha_k = 1$. Thus for $t > 0$ and all k, $0 \leq \alpha_k \leq 1$, or equivalently $\sigma_{up} + x_k^{up} \leq x_k \leq x_k^{up}$. This implies that an escort dynamic of the form

$$\dot{x}_k = \alpha_k(f_k(\mathbf{x}) - \bar{f}_\alpha(\mathbf{x})), \tag{5.10}$$

with (5.2) as definition of the weighted average payoff, will provide states $\mathbf{x} \in \Delta_{up}^K$, i.e. the simplex Δ_{up}^K is invariant under the ED of the form (5.10).

A similar reasoning applies if $\phi_k(x_k) = \beta_k$, and at $t = 0$, x_k is such that $\beta_k > 0$ for all k, then for $t > 0$ and all k, $0 \leq \beta_k \leq 1$, or equivalently $x_k^{lo} \leq x_k \leq \sigma_{lo} + x_k^{lo}$. Thus, an escort dynamic of the form

$$\dot{x}_k = \beta_k(f_k(\mathbf{x}) - \bar{f}_\beta(\mathbf{x})), \tag{5.11}$$

will provide states $\mathbf{x} \in \Delta_{lo}^K$, i.e. the simplex Δ_{lo}^K is invariant under the ED of the form (5.11).

Now let us evaluate the case where $\phi_k(x_k) = \alpha_k\beta_k$ is chosen as the escort function. If at $t = 0$, x_k is such that $\alpha_k\beta_k > 0$ for all k, then it is also true that both $\alpha_k > 0$

and $\beta_k > 0$ at $t = 0$. Given that the escort dynamic guarantees $\sum_{k=1}^{K} x_k = 1$ for $t > 0$, then from (5.9) and (5.8), $\sum_{k=1}^{K} \alpha_k = 1$ and $\sum_{k=1}^{K} \beta_k = 1$ are also guaranteed for $t > 0$. Then both intervals $0 \leq \alpha_k \leq 1$ and $0 \leq \beta_k \leq 1$ are respected, or equivalently both $\sigma_{up} + x_k^{up} \leq x_k \leq x_k^{up}$ and $x_k^{lo} \leq x_k \leq \sigma_{lo} + x_k^{lo}$ are respected. This last statement is only valid if the intersection exists. Thus, an ED of the form

$$\dot{x}_k = \alpha_k \beta_k (f_k(\mathbf{x}) - \bar{f}_{\alpha\beta}(\mathbf{x})), \tag{5.12}$$

will provide states $\mathbf{x} \in \Delta_{lo}^K \cap \Delta_{up}^K$. This result is remarkable and it allows the development of the decentralized approach followed in this chapter.

Taking into account the nature of the proposed escort functions, it is possible to propose candidates for Lyapunov functions that allow to prove the stability of the dynamics at an equilibrium point $\hat{\mathbf{x}}$. In the definitions of [Har11], escort functions were imposed to be positive and monotonically increasing. For the intersection escort functions α_k and β_k are positive, but one of them decreases as the other increases. As a result the product is not monotonically increasing but is still positive. It is possible to prove that the following divergence-like function,

$$L(\mathbf{x}) = \sum_{k=1}^{K} \frac{\sigma_{up}\sigma_{lo}}{x_k^{up} - x_k^{lo}} \left\{ (\hat{x}_k - x_k^{up}) \log\left(\frac{\hat{x}_k - x_k^{up}}{x_k - x_k^{up}} \right) - (\hat{x}_k - x_k^{lo}) \log\left(\frac{\hat{x}_k - x_k^{lo}}{x_k - x_k^{lo}} \right) \right\},$$

is a Lyapunov function for the ED with the proposed intersection escort functions, and equilibrium state $\hat{\mathbf{x}}$.

5.5 Examples

To differentiate the escort gradient from the shahshahani gradient of the previous chapter, let us consider some examples. It should be mentioned that the escort gradient is a generalization of the shahshahani gradient. In fact, the escort dynamics generalizes other well-known evolutionary game theory dynamics like the replicator and mixed strategist dyamics (Shahshahani gradient), and the projection dynamics [Har11, Har09, SDL08].

A potential function is defined as $F(\mathbf{x}) = -(x_1 - 13/20)^2 - (x_2)^2 - (x_3 - 17/20)^2$, for $\mathbf{x} \in \mathbb{R}^3$. This function has a global maximum in $\hat{\mathbf{x}}^+ = [13/20, 0, 17/20]^T$, which is a point in the plane $\{x_1 + x_2 + x_3 = 1.5\} \in \mathbb{R}^3$. However, if the feasible region is constrained to points $\mathbf{x} \in \{x_1 + x_2 + x_3 = 1\}$, then the feasible maximum is $\hat{\mathbf{x}}^* = [29/60, -10/60, 41/60]^T$. Furthermore, if the feasible region is even more constrained to points $\mathbf{x} \in \Delta^3 = \{x_1, x_2, x_3 \geq 0, x_1 + x_2 + x_3 = 1\}$, then the feasible maximum is $\hat{\mathbf{x}} = [2/5, 0, 3/5]^T$.

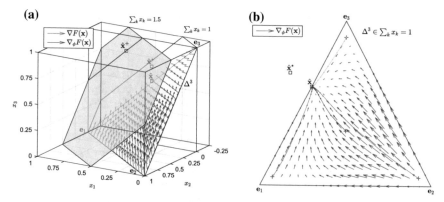

Fig. 5.5 Euclidean and Escort gradient vector fields (scaled) for an example of potential function with domain $\mathbf{x} \in \mathbb{R}^3$. The feasible maximum $\hat{\mathbf{x}}$ lies in the simplex Δ^3 while the unfeasible maxima $\hat{\mathbf{x}}^+$ and $\hat{\mathbf{x}}^*$ lie respectively in planes $\{x_1 + x_2 + x_3 = 1.5\} \in \mathbb{R}^3$, and $\{x_1 + x_2 + x_3 = 1\} \in \mathbb{R}^3$ ($\square$: feasible and unfeasible maxima). **a** Escort gradient with escort functions $\phi_k(x_k) = x_k$ (Shahshahani gradient). **b** Closer look on the plane $\{x_1 + x_2 + x_3 = 1\}$, the escort gradient vector field, and some trajectories converging to the feasible optimum $\hat{\mathbf{x}}$

Let us use this example to illustrate a comparison between the euclidean gradient and the escort gradient. For several points $\mathbf{x}$ inside the simplex Δ^3, Fig. 5.5a, c show the vector fields of both the euclidean and the escort (two types of escort functions) gradients as well as the global and feasible maxima, $\hat{\mathbf{x}}^+$, $\hat{\mathbf{x}}^*$, and $\hat{\mathbf{x}}$, for the proposed example. Besides, Fig. 5.5b, d show a more clear view of the simplex Δ^3 in the plane $\{x_1 + x_2 + x_3 = 1\}$ and the vector fields of the escort gradient in two cases for the same function $F(\mathbf{x})$ of the example. As it was mentioned in Chap. 4, the euclidean gradient vector $\nabla F(\mathbf{x})$ points in the direction of maximal increase of the potential function $F(\mathbf{x})$. Consequently it points to the plane $\{x_1 + x_2 + x_3 = 1.5\}$ which contains the unconstrained maximum $\hat{\mathbf{x}}^+$.

On the other hand, as it was explained with Eq. (5.3), the escort gradient vector $\nabla_\phi F(\mathbf{x})$ lies on the tangent space $T_{\mathbf{x}}\Delta^K$ of the simplex Δ^K, which in words means that both its tail (i.e. $\mathbf{x}$) and tip (i.e. $\mathbf{x} + \nabla_\phi F(\mathbf{x})$) lie on the plane $\{x_1 + x_2 + x_3 = 1\} \supset \Delta^3$ for this example. If the escort functions are chosen to be $\phi_k(x_k) = x_k$, then the escort gradient becomes the shashahani gradient. In this case, Fig. 5.5a, b show the vector field of the escort gradient $\nabla_\phi F(\mathbf{x})$ when it becomes the shahshahani gradient, for the proposed example. With these escort functions, Fig. 5.5b shows 4 trajectories starting from different points inside the simplex. It is possible to observe that all of them converge asymptotically to the feasible maximum $\hat{\mathbf{x}}$. Given that these escort functions satisfy $\phi_k(0) = 0$, then the trajectories are not allowed to leave the simplex Δ^3.

5.5.1 Examples with the Proposed Intersection Escort Functions

As it was mentioned in Sect. 5.4, in this chapter the purpose of the escort functions is further exploited for more general optimization scenarios, specially for the problem of decentralized management of PEVs load. A first look on the ED compared to the Shahshahani gradient does not allow to draw conclusions on the difference between these approaches. However, the examples on this subsection are devoted to explain the propositions of this chapter in terms of escort functions. More specifically, the examples in this subsection show how escrt functions can be used to achieve similar results to those of the last chapter and even have a much more flexible handling of the constraints involved in the optimization problems. As it will be shown in the following sections, the approach proposed in this chapter is much more flexible than those previous chapters. It is capable of handling a much wider range of constraints as it well be discussed.

To illustrate the purpose of the *intersection escort functions* proposed in this chapter, let us introduce some examples in three and four dimensions. Taking into account the same potential function $F(\mathbf{x}) = -(x_1 - 13/20)^2 - (x_2)^2 - (x_3 - 17/20)^2$, for $\mathbf{x} \in \mathbb{R}^3$, with global maximum at $\hat{\mathbf{x}}^+ = [13/20, 0, 17/20]^T$, which is a point in the plane $\{x_1 + x_2 + x_3 = 1.5\} \in \mathbb{R}^3$. Again, if the feasible region is constrained to points $\mathbf{x} \in \{x_1 + x_2 + x_3 = 1\}$, then the feasible maximum is $\hat{\mathbf{x}}^* = [29/60, -10/60, 41/60]^T$.

Now, let us consider lower limits on each dimension, so the feasible region is now, $\mathbf{x} \in \{x_1 \geq 0.075, x_2 \geq -0.05, x_3 \geq 0.1, x_1 + x_2 + x_3 = 1\}$. Given these lower constraints, the resulting matrix $\mathbf{C}_{lo}$ computed with (5.7) is,

$$
\mathbf{C}_{lo} = \mathbf{X}_{lo} + \sigma_{lo}\mathbf{I} = \begin{bmatrix} 0.075 & 0.075 & 0.075 \\ -0.05 & -0.05 & -0.05 \\ 0.1 & 0.1 & 0.1 \end{bmatrix} + 0.875 \begin{bmatrix} 1 & 0 & 0 \\ 0 & 1 & 0 \\ 0 & 0 & 1 \end{bmatrix} = \begin{bmatrix} 0.95 & 0.075 & 0.075 \\ -0.05 & 0.825 & -0.05 \\ 0.1 & 0.1 & 0.975 \end{bmatrix},
$$

while the resulting escort functions given by (5.8) are,

$$
\beta_1 = \frac{x_1 - 0.075}{0.875}, \qquad \beta_2 = \frac{x_2 + 0.05}{0.875}, \qquad \beta_3 = \frac{x_3 - 0.1}{0.875}.
$$

As it was explained, the columns of $\mathbf{C}_{lo}$ are the vertices of the simplex Δ_{lo}^K. For this example, this simplex is shown on Fig. 5.6a, b. In this case, the feasible optimum is $\hat{\mathbf{x}} = [0.425, -0.05, 0.625]^T$. Figure 5.6a shows the comparison between the euclidean gradient and the escort gradient with the computed escort functions. As it was explained before, the escort gradient is a vector with its tip and tail on the plane $\sum_k x_k = 1$, so the trajectories followed by the ED are always on this plane. Besides, the computed escort functions β_k are positive inside the simplex Δ_{lo}^3, and become zero at the predefined limits, so the trajectories are confined inside this simplex. Figure 5.6b shows four sample trajectories arriving to the feasible optimum, as well as a more detailed view of the escort gradient vector field (with escort functions β_k) on the plane $x_1 + x_2 + x_3 = 1$.

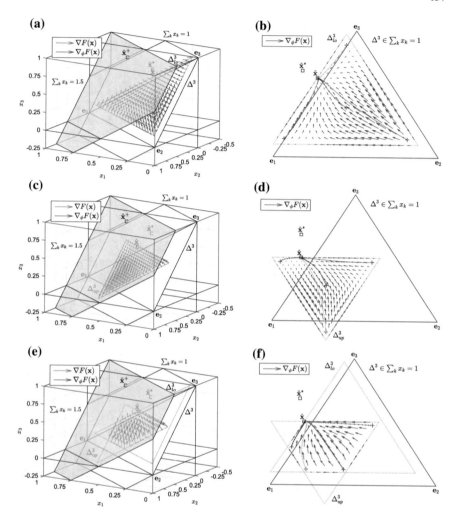

Fig. 5.6 Euclidean and Escort gradient vector fields (scaled) for an example of potential function with domain $\mathbf{x} \in \mathbb{R}^3$. The feasible maximum $\hat{\mathbf{x}}$ lies in the simplex Δ^3 while the unfeasible maxima $\hat{\mathbf{x}}^+$ and $\hat{\mathbf{x}}^*$ lie respectively in the planes $\{x_1 + x_2 + x_3 = 1.5\} \in \mathbb{R}^3$, and $\{x_1 + x_2 + x_3 = 1\} \in \mathbb{R}^3$ ($\square$: feasible and unfeasible maxima). **a** Escort gradient with proposed escort functions $\phi_k(x_k) = \beta_k$. **b** Closer look on the plane $\{x_1 + x_2 + x_3 = 1\}$, the escort gradient vector field, and some trajectories converging to the feasible optimum on the lower constraints simplex Δ_{lo}^3. **c–f** Similar figures for escort functions $\phi_k(x_k) = \alpha_k$, and $\phi_k(x_k) = \alpha_k \beta_k$. Trajectories converging to feasible optima in the respective regions Δ_{up}^3, and $\Delta_{lo}^3 \cap \Delta_{up}^3$

Let us consider now upper constraints on each dimension, so the feasible region is now, $\mathbf{x} \in \{x_1 \leq 0.75, x_2 \leq 0.4, x_3 \leq 0.5, x_1 + x_2 + x_3 = 1\}$. Given these constraints, the resulting matrix $\mathbf{C}_{up}$ computed with (5.7) is,

$$\mathbf{C}_{up} = \mathbf{X}_{up} + \sigma_{up}\mathbf{I} = \begin{bmatrix} 0.75 & 0.75 & 0.75 \\ 0.4 & 0.4 & 0.4 \\ 0.5 & 0.5 & 0.5 \end{bmatrix} - 0.65 \begin{bmatrix} 1 & 0 & 0 \\ 0 & 1 & 0 \\ 0 & 0 & 1 \end{bmatrix} = \begin{bmatrix} 0.1 & 0.75 & 0.75 \\ 0.4 & -0.25 & 0.4 \\ 0.5 & 0.5 & -0.15 \end{bmatrix},$$

while the resulting escort functions given by (5.8) are,

$$\alpha_1 = \frac{-x_1 + 0.75}{0.65}, \qquad \alpha_2 = \frac{-x_2 + 0.4}{0.65}, \qquad \alpha_3 = \frac{-x_3 + 0.5}{0.65}.$$

Now, the columns of $\mathbf{C}_{up}$ are the vertices of the simplex Δ_{up}^3 as it is is shown on Fig. 5.6c, d. The feasible optimum is now $\hat{\mathbf{x}} = [0.575, -0.075, 0.5]^T$. Figure 5.6c shows the comparison between the euclidean gradient and the escort gradient with the computed escort functions α_k. In this case, the trajectories followed by the ED are confined inside Δ_{up}^3 given that the computed escort functions α_k are positive inside the simplex Δ_{up}^3, and become zero at the predefined upper limits. Figure 5.6d shows other four sample trajectories arriving to the new feasible optimum, as well as a more detailed view of the escort gradient vector field (with escort functions α_k) on the plane $x_1 + x_2 + x_3 = 1$.

Let us now consider both types of escort functions and combine them as in Eq. (5.12). This results in the following escort functions,

$$\phi_1 = \frac{(x_1 - 0.075)(0.75 - x_1)}{0.56875}, \quad \phi_2 = \frac{(x_2 + 0.05)(0.4 - x_2)}{0.56875}, \quad \phi_3 = \frac{(x_3 - 0.1)(0.5 - x_3)}{0.56875},$$

which are positive valued for points $\mathbf{x} \in \{0.075 \leq x_1 \leq 0.75, -0.05 \leq x_2 \leq 0.4, 0.1 \leq x_3 \leq 0.5\}$. As it was explained in Sect. 5.4.2, if $\{\phi_1, \phi_2, \phi_3\}$ are positive, then it is true that $\{\alpha_1, \alpha_2, \alpha_3, \beta_1, \beta_2, \beta_3\}$ are positive as well. Given that the ED guarantees $\{x_1 + x_2 + x_3 = 1\}$ always, then,

$$\alpha_1 + \alpha_2 + \alpha_3 = \frac{-x_1 + 0.75 - x_2 + 0.4 - x_3 + 0.5}{0.65} = \frac{(0.75 + 0.4 + 0.5) - 1}{0.65} = 1,$$

$$\beta_1 + \beta_2 + \beta_3 = \frac{x_1 - 0.075 + x_2 + 0.05 + x_3 - 0.1}{0.875} = \frac{(-0.075 + 0.05 - 0.1) + 1}{0.875} = 1,$$

which are consequently always satisfied as well. Then, any trajectory followed by the ED with escort functions $\{\phi_1, \phi_2, \phi_3\}$, satisfies both sets $\{\alpha_1, \alpha_2, \alpha_3 \geq 0, \alpha_1 + \alpha_2 + \alpha_3 = 1\}$ and $\{\beta_1, \beta_2, \beta_3 \geq 0, \beta_1 + \beta_2 + \beta_3 = 1\}$. Consequently, all the escort functions also have upper limits at $\{\alpha_1, \alpha_2, \alpha_3, \beta_1, \beta_2, \beta_3 \leq 1\}$ and the intersections among the following intervals are respected,

$$0 \le \alpha_1 \le 1, \qquad 0 \le \alpha_2 \le 1, \qquad 0 \le \alpha_3 \le 1,$$

$$0 \le \frac{-x_1 + 0.75}{0.65} \le 1, \qquad 0 \le \frac{-x_2 + 0.4}{0.65} \le 1, \qquad 0 \le \frac{-x_3 + 0.5}{0.65} \le 1,$$

$$0.1 \le x_1 \le 0.75, \qquad -0.25 \le x_2 \le 0.4, \qquad -0.15 \le x_3 \le 0.5,$$

$$0 \le \beta_1 \le 1, \qquad 0 \le \beta_2 \le 1, \qquad 0 \le \beta_3 \le 1,$$

$$0 \le \frac{x_1 - 0.075}{0.875} \le 1, \qquad 0 \le \frac{x_2 + 0.05}{0.875} \le 1, \qquad 0 \le \frac{x_3 - 0.1}{0.875} \le 1,$$

$$0.075 \le x_1 \le 0.95, \qquad -0.05 \le x_2 \le 0.825, \qquad 0.1 \le x_3 \le 0.975.$$

This can be observed in Fig. 5.6e, f where the escort gradient vector field is shown for the escort functions $\{\phi_1, \phi_2, \phi_3\}$. Now, the feasible region is $\mathbf{x} \in \{0.1 \le x_1 \le 0.75, -0.05 \le x_2 \le 0.4, 0.1 \le x_3 \le 0.5, x_1 + x_2 + x_3 = 1\}$, and the feasible optimum is $\hat{\mathbf{x}} = [0.55, -0.05, 0.5]^T$. As well as for the other types of escort functions, Fig. 5.6f shows shows four sample trajectories arriving to the feasible optimum, and a more detailed view of the escort gradient vector field (with escort functions ϕ_k) on the plane $x_1 + x_2 + x_3 = 1$.

5.6 The ED Approach for PEV Load Management

Let us consider a matrix π, whose elements $\pi_{k,m}$ correspond to the total energy consumption rate (active power) on phases $m = \{1, 2, 3\}$ of a distribution system transformer at times $k = \{1, 2, \ldots, K\}$. Accordingly, we can consider a similar matrix ρ, whose elements $\rho_{k,m}$ correspond to the total reactive power provided by each of the phases of the same transformer. Elements $\pi_{k,m}$ and $\rho_{k,m}$ are defined as follows,

$$\pi_{k,m} = p_{k,m} + \sum_{i=1}^{J_k} x_{k,m}^i / \tau, \qquad \rho_{k,m} = q_{k,m} + \sum_{i=1}^{J_k} y_{k,m}^i,$$

where $p_{k,m}$ and $q_{k,m}$ correspond respectively to the forecasted active and reactive powers for each of the transformer's phases without considering PEVs. Parameter K is the number of time slots in the forecasted horizon, where each time slot has a duration τ (in hours or fractions). Recalling Sect. 5.2, portions of *sedentary* populations on each phase m at a time time slot k represent parameters $p_{k,m}$ in the energy analogy. For the reactive power analogy, initially available *hosting capacities* per phase and time slot represent parameters $q_{k,m}$.

Parameters $x_{k,m}^i / \tau$ and $y_{k,m}^i$ respectively correspond to the active and reactive powers from PEVs per phase, and J_k is the number of connected PEVs at time k.

Recalling the analogies Sect. 5.2 again, parameters $x^i_{k,m}/\tau$ and $y^i_{k,m}$ are represented by portions of populations[4] allocated in territories (k, m).

5.6.1   Formal Definitions for Populations Representing PEV Quantities

A PEV i must consume an amount E^i of energy to fully charge its battery in K^i time slots. This duration is defined or estimated by the owner who can be as restrictive required (taking into account the charger's limits). In the case of three-phase chargers, the consumption can be distributed among time slots and phases. The amount of energy E^i defines the energy population size as,

$$E^i = soc^i_d - soc^i_0, \tag{5.13}$$

where soc^i_d is the desired state of charge (Wh) at the end of the time window defined by the owner, and soc^i_0 is the initial state of charge. Thus, for a PEV, this parameter defines the hyper-plane where its energy population distribution lie (before normalization),

$$\sum_{k=1}^{K^i} \sum_{m=1}^{3} x^i_{k,m} = soc^i_d - soc^i_0. \tag{5.14}$$

Then, the simplices $\Delta^K_{lo,i}$ and $\Delta^K_{up,i}$ for the energy population of PEV i are defined by hyper-plane (5.14) and the following linear inequalities (half-spaces),

$$x^i_{k,m} \leq \overline{soc}^i - \left(soc^i_0 + \sum_{\omega=1}^{\Omega} \sum_{m=1}^{3} x^i_{\omega,m} - x^i_{k,m} \right), \tag{5.15a}$$

$$x^i_{k,m} \geq \underline{soc}^i - \left(soc^i_0 + \sum_{\omega=1}^{\Omega} \sum_{m=1}^{3} x^i_{\omega,m} - x^i_{k,m} \right), \tag{5.15b}$$

$$\forall \Omega = \{1, 2, \ldots, K^i\}, \ \forall k = \{1, 2, \ldots, \Omega\}, \ \forall m = \{1, 2, 3\}$$
$$-\tau \overline{p}^i \leq x^i_{k,m} \leq \tau \overline{p}^i, \tag{5.15c}$$

$$x^i_{k,m} \leq \tau(\overline{L} - \pi_{k,m}) + x^i_{k,m}, \tag{5.15d}$$
$$\forall k = \{1, 2, \ldots, K^i\}, \ \forall m = \{1, 2, 3\},$$

---

[4]*Nomad populations* for the energy analogy.

where $\overline{p}^i$ is the nominal power of the charger per phase, and $\overline{L}$ is the limit load of the transformer per phase. Constraints (5.15a) and (5.15b) represent the accumulated state of charge at time slot Ω which cannot exceed lower and upper limits ($\underline{soc}^i$ and $\overline{soc}^i$) designed to protect the battery state of health. Constraint (5.15c) represents the limits of energy consumption/injection given the charger's nominal power and the step size τ. Finally, constraint (5.15d) represents the limits of energy consumption given the nominal limits of the transformer.

Simplex $\Delta_{up,i}^K$ is defined by the intersection of upper constraints in (5.15a), (5.15c) and (5.15d), while simplex $\Delta_{lo,i}^K$ is defined by the intersection of lower constraints in (5.15b) and (5.15c).

On the other hand, for PEV i, parameter Q^i is the size of its population representing *reactive power*. It depends on the energy population distribution and the charger's nominal charging rate $\overline{p}^i$, and it is given by,

$$Q^i = \sum_{k=1}^{K^i} \sum_{m=1}^{3} |\overline{q}_{k,m}^i| = \sum_{k=1}^{K^i} \sum_{m=1}^{3} \sqrt{(\overline{p}^i)^2 - (x_{k,m}^i/\tau)^2},$$

where $\overline{q}_{k,m}^i = \pm\sqrt{(\overline{p}^i)^2 - (x_{k,m}^i/\tau)^2}$ is the upper and lower limit of reactive power per phase m at time slot k. As it is expected, during the connection time $k = \{1, 2, \ldots, K^i\}$, the PEV is only able to supply at most (or at least) the size of its reactive power population Q^i, i.e.,

$$-Q^i \le \sum_{k=1}^{K^i} \sum_{m=1}^{3} y_{k,m}^i \le Q^i,$$

To maintain the hyper-plane notation, let us introduce a slack variable $-Q^i \le s^i \le Q^i$ representing the portion of of the reactive power population that is not allocated to any phase at any time slot. Then, the hyper-plane where the reactive power population distribution lies is defined by,

$$\sum_{k=1}^{K^i} \sum_{m=1}^{3} y_{k,m}^i + s^i = 0, , \tag{5.16}$$

and the corresponding simplices $\Delta_{lo,i}^K$ and $\Delta_{up,i}^K$ (before normalization) are defined by this same hyper-plane and the following linear constraints (half-spaces),

$$-\overline{q}_{k,m}^i \le y_{k,m}^i \le \overline{q}_{k,m}^i, \tag{5.17a}$$

$$\forall k = \{1, 2, \ldots, K^i\}, \ \forall m = \{1, 2, 3\},$$
$$-Q^i \le s^i \le Q^i. \tag{5.17b}$$

5.6.1.1   Adaptations for Single-Phase Chargers

For single-phase chargers the number of available environments is reduced but the definitions do not change. For instance, for a PEV connected to phase m, the hyperplane where its energy population lies become,

$$\sum_{k=1}^{K^i} x_{k,m}^i = soc_d^i - soc_0^i, \tag{5.18}$$

and the simplices $\Delta_{lo,i}^K$ and $\Delta_{up,i}^K$ (before normalization), are defined by the hyperplane and the following half-spaces,

$$x_{k,m}^i \leq \overline{soc}^i - \left(soc_0^i + \sum_{\omega=1}^{\Omega} x_{\omega,m}^i - x_{k,m}^i \right), \tag{5.19a}$$

$$x_{k,m}^i \geq \underline{soc}^i - \left(soc_0^i + \sum_{\omega=1}^{\Omega} x_{\omega,m}^i - x_{k,m}^i \right), \tag{5.19b}$$

$$\forall \Omega = \{1, 2, \ldots, K^i\}, \ \forall k = \{1, 2, \ldots, \Omega\},$$

$$- \tau \overline{p}^i \leq x_{k,m}^i \leq \tau \overline{p}^i, \tag{5.19c}$$

$$x_{k,m}^i \leq \tau(\overline{L} - \pi_{k,m}) + x_{k,m}^i, \tag{5.19d}$$

$$\forall k = \{1, 2, \ldots, K^i\},$$

For the same PEV, the reactive power population distribution lies in the hyperplane,

$$\sum_{k=1}^{K^i} y_{k,m}^i + s^i = 0, \tag{5.20}$$

and the corresponding simplices $\Delta_{lo,i}^K$ and $\Delta_{up,i}^K$ are defined by the same hyper-plane and the following half-spaces,

$$- \overline{q}_{k,m}^i \leq y_{k,m}^i \leq \overline{q}_{k,m}^i, \tag{5.21a}$$

$$\forall k = \{1, 2, \ldots, K^i\},$$

$$- Q^i \leq s^i \leq Q^i, \tag{5.21b}$$

where the size of the population is defined as $Q^i = \sum_{k=1}^{K^i} \sqrt{(\overline{p}^i)^2 - (x_{k,m}^i/\tau)^2}$. As it may be observed, the single-phase case is a degenerate version of the three-phase case, considering only the concerned variables associated to the phase where the charger is connected.

It is important to clarify the origin of constraints (5.15a), and (5.15b) for three-phase chargers, and (5.19a) (5.19b) for single-phase chargers. These expressions result from the manipulation of constraints imposed to the partial states of charge. For instance, let us consider a PEV connected to a single-phase charger for a time window $k = \{1, 2, 3\}$ ($\tau = 1h$), and neglect the phase m and vehicle i indices. In this case, the three partial states of charge are constrained as follows,

$$\underline{soc} \le soc_1 \le \overline{soc} \Rightarrow \underline{soc} \le soc_0 + x_1 \le \overline{soc},$$
$$\underline{soc} \le soc_2 \le \overline{soc} \Rightarrow \underline{soc} \le soc_0 + x_1 + x_2 \le \overline{soc},$$
$$\underline{soc} \le soc_3 \le \overline{soc} \Rightarrow \underline{soc} \le soc_0 + x_1 + x_2 + x_3 \le \overline{soc}.$$

Clearing for each of the variables of interest (x_1, x_2, and x_3) results in constraints,

$$\underline{soc} - soc_0 \le x_1 \le \overline{soc} - soc_0,$$
$$\underline{soc} - soc_0 - x_2 \le x_1 \le \overline{soc} - soc_0 - x_2,$$
$$\underline{soc} - soc_0 - x_2 - x_3 \le x_1 \le \overline{soc} - soc_0 - x_2 - x_3,$$
$$\underline{soc} - soc_0 - x_1 \le x_2 \le \overline{soc} - soc_0 - x_1,$$
$$\underline{soc} - soc_0 - x_1 - x_3 \le x_2 \le \overline{soc} - soc_0 - x_1 - x_3,$$
$$\underline{soc} - soc_0 - x_1 - x_2 \le x_3 \le \overline{soc} - soc_0 - x_1 - x_2,$$

These expressions represent the effect of constrained partial states of charge soc_k^i over the variables of charge x_k^i at each slot of time k.

5.6.2 Payoff Function Definitions

Payoff functions are proposed such that the goals of the utility grid and the goals of the PEV owners can be handled simultaneously. For this reason, a commitment factor μ^i and a smoothing factor η are proposed. The first factor, can be controlled by PEV owners and is proposed to give a level of choice to the owners. It is proposed in such a way that the the utility grid manager offers a motivation or incentive given the owner's commitment to the grid. These factors are defined as,

$$\underline{\mu} \le \mu^i \le 1, \quad 0 \le \eta \le 1,$$

where $0 \le \underline{\mu} \le 1$ is the minimum allowed, and 1 is the maximum level of commitment. The smoothing parameter η is defined by the utility grid manager to reduce drastic changes in variables from one time step to the next.

The following monotonically decreasing payoff functions are proposed such that the potential functions are quadratic, strictly concave on all the controllable parameters. The zero crossing of these functions is defined by the level of commitment and the smoothing parameter. For *energy* populations, the proposed payoff functions per pure strategy (i.e. per phase at a given time step) are,

$$f_{k,m}^i(x_{k,m}^i) = -(1-\mu^i)(x_{k,m}^i - x_{k,m}^{i*})/\tau \tag{5.22}$$
$$-\mu^i\left[\eta\pi_{k,m} + (1-\eta)(2\pi_{k,m} - \pi_{k-1,m} - \pi_{k+1,m})\right],$$

where $x_{k,m}^{i*}$ is the owner's preferred reference for the charging rate (i.e. the reference for power consumption of the charger at time k and phase m). The interests of the utility grid manager in this case are focused in the balance of the active power drawn from each of the transformer's phases and the flattening of the corresponding active power profiles. On the other hand, the owners interests depend on their behavior regarding its mobility needs, and the incentive from the utility manager for their commitment.

Similarly, for *reactive power* populations the proposed functions are,

$$g_{k,m}^i(y_{k,m}^i) = -(1-\mu^i)y_{k,m}^i \tag{5.23a}$$
$$-\mu^i\left[\eta\rho_{k,m} + (1-\eta)(2\rho_{k,m} - \rho_{k-1,m} - \rho_{k+1,m})\right]$$
$$g^i(s^i) = 0. \tag{5.23b}$$

Here, payoff functions (5.23b) are zero valued because the slack variables do not have any influence on the average payoff of the populations. Regarding reactive power, there is no special interest from the owner's point of view besides keeping it as close as possible to zero. This is the reason why terms with $(1-\mu^i)$ disappear in (5.23a).

5.6.2.1 Payoff Functions Definition for Single-Phase Chargers

For single-phase chargers, the definitions of payoff functions are similar. Nevertheless, an additional parameter v is included to balance between single-phase grid's interests and three-phase grid's interests. This parameter has values

$$0 \le v \le 1,$$

and it is included in the definitions of payoff functions as follows. For energy populations and single-phase chargers the proposed payoff functions are,

$$f_{k,m}^i(x_{k,m}^i) = -(1-\mu^i)(x_{k,m}^i - x_{k,m}^{i*})/\tau \tag{5.24}$$
$$-\mu^i v\left[\eta\pi_{k,m} + (1-\eta)\left(2\pi_{k,m} - \pi_{k-1,m} - \pi_{k+1,m}\right)\right]$$
$$-\mu^i(1-v)\left[\eta\sum_{m=1}^{3}\pi_{k,m} + (1-\eta)\sum_{m=1}^{3}(2\pi_{k,m} - \pi_{k-1,m} - \pi_{k+1,m})\right].$$

Similarly, for reactive power populations and single-phase chargers the proposed functions are,

$$g^i_{k,m}(y^i_{k,m}) = -(1 - \mu^i)y^i_{k,m} \tag{5.25a}$$
$$- \mu^i \upsilon \left[\eta \rho_{k,m} + (1 - \eta)(2\rho_{k,m} - \rho_{k-1,m} - \rho_{k+1,m}) \right]$$
$$- \mu^i (1 - \upsilon) \left[\eta \sum_{m=1}^{3} \rho_{k,m} + (1 - \eta) \sum_{m=1}^{3} (2\rho_{k,m} - \rho_{k-1,m} - \rho_{k+1,m}) \right]$$
$$g^i(s^i) = 0. \tag{5.25b}$$

The purpose of parameters μ^i, η, and υ is summarized in Fig. 5.7. The average value μ of parameters μ^i (defined by owners), and η (defined by the grid manager), have an effect in the final output profiles of total and single phase active and reactive power profiles. If $\mu = 0$, the smoothing/flattening/balancing objectives are neglected, while payoff functions give value only to local references of load distribution. On the other hand, if $\eta = 0$ and $\mu > 0$, payoff functions are tuned such that the smoothing objective is the priority, while the flattening/balancing objectives are neglected. The last extreme case occurs when $\mu > 0$, and $\eta = 1$. This tunes the payoff functions such that the flattening/balancing objectives are the only priority, while the smoothing objective is neglected, making final profiles become piece-wise balanced and flat (according to the presence of PEVs). In all cases, the utility grid manager has both direct and indirect control over these parameter: it defines η by the trade-off of its grid objectives, while it influences the mean value of μ with social or economic incentives to vehicle owners.

In [Ova+17], instead of using individual parameters μ^i for each local ED routine, the average value μ is considered to be supplied by the aggregator given all the μ^i values chosen by the PEV owners. Using the Matlab scripts at the end of this chapter,

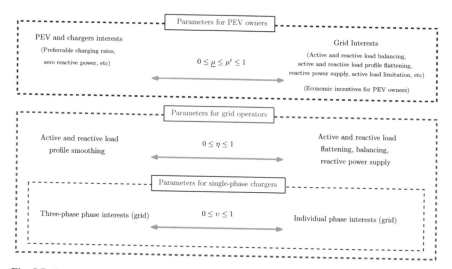

Fig. 5.7 Summary of the role of parameters μ^i, η, and υ and the effect they have on the proposed integral PEV load management approach

readers are encouraged to do sweeps over the range of these parameter values and observe their effect. Readers are also encouraged to modify the payoff functions to test different alternatives.

5.7 The ED Based Distributed Management Loop

Based on the definitions of the section before and the ED discretization, Fig. 5.8 summarizes the proposed algorithm. Similar to Chap. 4, here an aggregator is in charge of managing the interaction among PEVs following a Best Reply dynamics [HS98]. It receives optimal distributions ($\hat{\mathbf{x}}^i$ and $\hat{\mathbf{y}}^i$) from PEVs, it aggregates these profiles to the forecasted active and reactive power profiles, and redistributes them to PEVs again. Thus, the aggregator makes the new information available to PEVs under the transformer. It is also in charge of providing and updating the forecasts of the consumtion on each phase of the MV-LV transformer using historic data [Bas+15, PS13]. Thus, it is constantly receiving and redistributing updated information as it is described in Fig. 5.8.

The aggregator manages when each PEV is able to update its profiles, and follows the order of arrival of PEVs. Once PEVs have updated their profiles, a new round of information exchanges starts. Despite of these asynchronous exchanges, PEVs fairly share the available resources and their contribution to the grid.

PEVs locally manage their distributions $\mathbf{x}^i$ and $\mathbf{y}^i$ given the owner's requirements of: time (K^i), state of charge (soc_d^i), and commitment to the grid (μ^i). Each time step k has a regular length τ defined by the forecasts time step size (in hours or fractions). However a vehicle may arrive or leave in the middle of one period. Thus for some PEVs, some initial and final time slots ($k = 1$ and $k = K^i$) may be $0 < \tau_k^i \leq \tau$.

Similar to the MSD approach on Chap. 4, a variance stopping criterion is defined by,

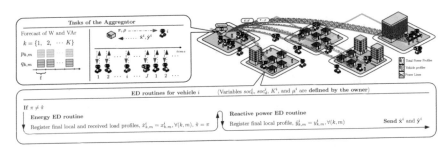

Fig. 5.8 Synergies and interactions in the proposed ED algorithm. The routines followed by each PEV locally, are based on the discretization of the ED equation for energy and reactive power populations. © [2017] IEEE. Reprinted, with permission, from [Ova+17]

$$\sum_{k=1}^{K} \left(\phi_k(x_k) \left(f_k(\mathbf{x}) - \sum_{k=1}^{K} \frac{\phi_k(x_k)f_k(\mathbf{x})}{\Phi(\mathbf{x})} \right)^2 \right) < \epsilon, \tag{5.26}$$

which is established since the change rate of the potential function $F(\mathbf{x})$ is its directional derivative along $\dot{\mathbf{x}}$, i.e. $\dot{F}(\mathbf{x}) = D_{\mathbf{x}}F(\dot{\mathbf{x}})$ [Har11]. By applying (5.4), it can be further developed until (5.26).

Parameter ϵ is defined according to the desired precision, however, the second stopping criteria is defined by a reduced amount of iterations. The local ED routines for energy and reactive power are provided as Matlab scripts in Sect. 5.9 at the end of this chapter. It includes step by step descriptions aiming to help the reader to understand how it behaves. It is also written and described to help the reader to re-program it other platforms according to her/his needs. Again, for the application on PEV scheduling routines, a backtracking line-search subroutine, similar to that described in [BV04], is included to have adaptive step sizes and increase the routine's convergence speed. In this case, the direction of descent employed in the backtracking line-search is provided by the ED gradient. It is up to the reader to use this subroutine or not, and test the differences in terms of performance.

It should be noticed that each PEV updates its active and reactive power profiles while the others wait with their profiles fixed. Once a PEV has fixed and sent its profiles, the others will have also their turns to update their own profiles. Thus, if a PEV reaches its optimal profiles in its turn, it will be forced to change them in the next turn because the conditions for the previously reached optima are no longer the same (i.e., they become suboptimal).

5.8 Numerical Examples

In this section, the proposed approach is tested and the results are separated in two parts. The first part is divided in 3 examples explaining how the approach behaves using single-phase, three-phase and both type of chargers, under simplified conditions. The second part presents an scenario under realistic conditions using both type of chargers. This examples are proposed in order to highlight most of the details of the ED approach. Nevertheless, additional scenarios can be considered given the variety of parameters to adjust (scenario conditions, parameters μ^i, η, and ν, types of chargers, etc.). We encourage the reader to formulate new scenarios and modify the provided scripts at the end of this chapter to test the outcome of multiple combinations of these parameters.

Table 5.1 Summary of Example 1: three-phase chargers only

Item	Description
Number of vehicles	6 PEVs
Chargers	All of them 3 kW/phase
Batteries	All of them 20 kWh
Initial states of charge	$soc_0^i = 55\%$
Desired states of charge	$soc_d^i = 80\%$
Utility grid manager	Service parameter fixed at $\eta = 0.5$
Trade-off factor	$\mu^i = 0.8$ for all PEVs
Time period	24 half hour steps ($K^i = 24$) between 05 and 17 h (disconnection at 17 h)

5.8.1 Example 1: Only Three-Phase Chargers

Let us consider 6 PEVs sharing similar conditions: connection time at 05 h, departure time at 17 h, battery capacities of 20 kWh, initial states of charge soc_0^i of 55%, desired soc_d^i of 80%, chargers power limits of ± 3 kW. On the other hand, time steps are half hours ($K^i = 24$ for all PEVs), and additional parameters are fixed as $\eta = 0.5$ defined by the grid manager, and $\mu^i = 0.8$ for all PEVs. A summary of these assumptions is included on Table 5.1.

Figure 5.9 shows the most interesting variables and results for this test, applying 50 rounds of PEVs-Aggregator information exchanges. Let us take a look on Fig. 5.9b. All the 6 PEVs reach the same apparent power distribution, as well as the same active and reactive power profiles for each of their three phases. This implies that they share the available active and reactive power resources evenly to reach their desired states of charge, and share the supply of reactive per phase. Fairness can also be observed in the evolution of the states of charge through the connection time, which is identical for the 6 PEVs. Figure 5.9c shows that both total active and reactive power profiles for each of the three phases are balanced and flattened in a smooth way throughout the 12 hour charging period. Besides, the forecasted demand of reactive power of each phase is almost completely supplied by PEVs.

The evolution of the distribution of apparent power on each phase for the 6 connected PEVs can be observed in Fig. 5.9a. Here, each vertex of the polygons represents one of the 24 half hour slots of the connection time. Six trajectories are shown within each phase's polygon. Points in these trajectories are convex linear combinations of the vertices of the polygon, with coefficients given by the normalized distribution of apparent power on each phase. This diagrams are useful to check how the apparent power distribution per PEV per phase evolves with its interaction with all other connected PEVs. All of them have uniform initial distributions of apparent power per phase, and with the evolution driven by the ED approach, all of them converge to a common final distribution per phase. These final distributions correspond to the three apparent power profiles of Fig. 5.9b.

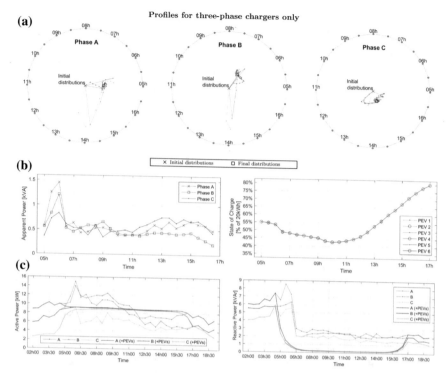

Fig. 5.9 6 PEVs connected to three-phase chargers under similar conditions. **a** evolution of the 6 normalized (scaled) distributions of apparent power per phase. **b** Final apparent power distributions, and state of charge profiles reached. **c** Forecast and final active and reactive power profiles per phase. © [2017] IEEE. Reprinted, with permission, from [Ova+17]

5.8.2 Example 2: Only Single-Phase Chargers

A similar test is performed with only single-phase chargers. Let us consider again 6 PEVs distributed over the three phases (2 per phase). They also share similar conditions: connection time at 05h, departure time at 17h, battery capacity of 20 kWh, soc_0^i of 55%, soc_d^i of 80%, chargers power limits of ± 3 kW. Time steps are half hours ($K^i = 24$ for all PEVs), and additional parameters are all fixed as $\eta = 0.5$ defined by the grid manager, and $\mu^i = 0.8$ for all PEVs. To check the effect on resource sharing, three values are considered for the third parameter v (defined by the grid manager as well). A summary of these assumptions can be found on Table 5.2.

Figure 5.10 shows the most interesting variables and results for this example, when the grid service parameter is fixed at $v = 1$. This parameter value that PEVs consider the smoothing/flattening objectives of the grid only for the phase where they are connected (See Fig. 5.7). In other words, PEVs will only consider the base forecast of active and reactive power of the phase where they are connected. The first thing to highlight is the fact that again the available resources are fairly shared

Table 5.2 Summary of Example 2: single-phase chargers only

Item	Description
Number of vehicles	6 PEVs
Chargers	All of them 3 kW
Batteries	All of them 20 kWh
Initial states of charge	$soc_0^i = 55\%$
Desired states of charge	$soc_d^i = 80\%$
Utility grid manager	Service parameter fixed at $\eta = 0.5$
Trade-off factor	$\mu^i = 0.8$ for all PEVs
Time period	24 half hour steps ($K^i = 24$) between 05 and 17 h (disconnection at 17 h)

among PEVs. However, this time PEVs observe phases as separated systems. The evolution of apparent power distributions per phase for the 6 PEVs can be observed in Fig. 5.10a. All of them have uniform initial distributions of apparent power on each phase, and with the evolution, all of them converge to a common final distribution on their corresponding phase. These final distributions correspond to the three apparent power profiles of Fig. 5.10b.

As it can be seen in the plots of Fig. 5.9b, pairs of PEVs on each phase reach the same apparent power distributions. For instance, the first pair of PEVs ($i = \{1, 2\}$), which are connected to phase A, both reach the same apparent power profiles labeled *Phase A*. Both of them will discharge their batteries more than PEVs connected in *phase B* and *phase C*, as it is observed in the state of charge profiles of the same figure. This occurs because at the time of connection (at 05h), Phase A has a more important forecasted demand than it is for phases B and C. Thus, these PEVs ($i = \{1, 2\}$) supply more energy to their phase to accomplish the smooth and flattening objective. The service provided by each pair of PEVs to their corresponding phase can be observed in Fig. 5.10c, for active and reactive power profiles per phase. Balance among phases is not achieved as it was for three-phase chargers on Fig. 5.9c. Fairness among PEVs is achieved but in the sense of separated ensembles defined per phase, not as a whole system. As it was with only three-phase chargers, desired states of charge are assured as well.

The last pair of diagrams on Fig. 5.9d, show the final total active and reactive power profiles achieved by PEVs with parameter $v = 1$. This plots are useful to compare with the case where this service parameter is fixed at the other extreme, $v = 0$. This is studied in the next part of this example.

Now, let us consider the same 6 PEVs, sharing similar conditions. However, this time parameter v (also defined by the grid manager as η) is fixed at $v = 0$. Figure 5.11 shows the most interesting outcomes obtained with this parameter value.

In this case, PEVs do not observe their phase as a separated system. Instead, PEVs consider only total active and reactive load forecasts to allocate their energy consumption and reactive power supply. The smoothing/flattening objectives of the

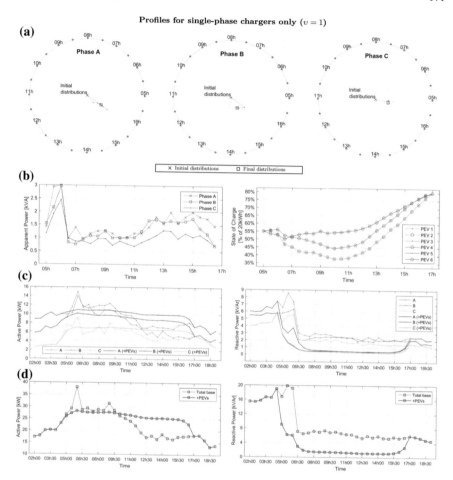

Fig. 5.10 PEVs connected to single-phase chargers ($v = 1$), by pairs on each phase, and sharing similar conditions. **a** evolution of the 6 normalized (scaled) apparent power distributions for the three phases. **b** Final apparent power distributions per phase, and state of charge profiles. **c** Forecast and final active and reactive power profiles obtained per phase. **d** Forecast and final total active and reactive power profiles

grid are taken into account not for each phase separately, but for the total three-phase system. As a consequence, each PEV updates its apparent (active and reactive) power distribution according to the same reference frame of all other PEVs, and not just of those connected to the same phase. The evolution of apparent power distributions per phase can be observed in Fig. 5.11a. All of them start with uniform distributions per phase, and with the evolution driven by the ED approach, all of them converge to a final common distribution neglecting the phase. This common final distribution corresponds to the apparent power profile shown on Fig. 5.10b. As a result, fairness

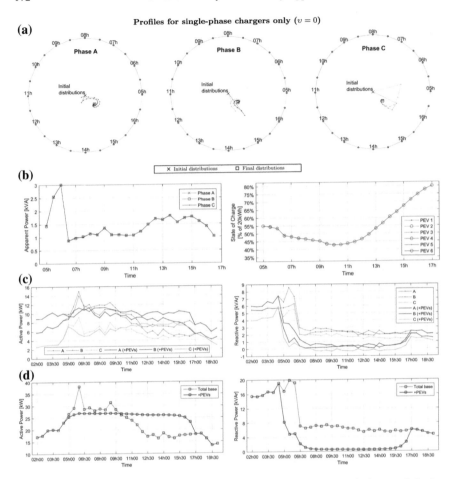

Fig. 5.11 6 PEVs connected to single-phase chargers ($v = 0$), by pairs on each phase, and sharing similar conditions. **a** Evolution of the 6 normalized (scaled) apparent power distributions for the three phases. **b** Final apparent power distributions per phase, and state of charge profiles. **c** Forecast and final active and reactive power profiles obtained per phase. **d** Forecast and final total active and reactive power profiles

is achieved among all the connected PEVs, regardless of their phase. This can also be noticed in the state of charge profiles of Fig. 5.11b.

However, fixing this service parameter at $v = 0$ has some consequences. In fat, it results in smoothing/flattening objectives for individual phases being neglected. Consequently, the outcome of final active and reactive power profiles per phase is ignored, as it can be noticed by comparing Fig. 5.11c with Fig. 5.10c. In this case, the attention of all PEVs is driven to provide services to the total active and reactive power profiles (the three-phase system), as it can be observed in Fig. 5.11d. This trade-up is acceptable since the main idea of the proposed approach is to provide a service

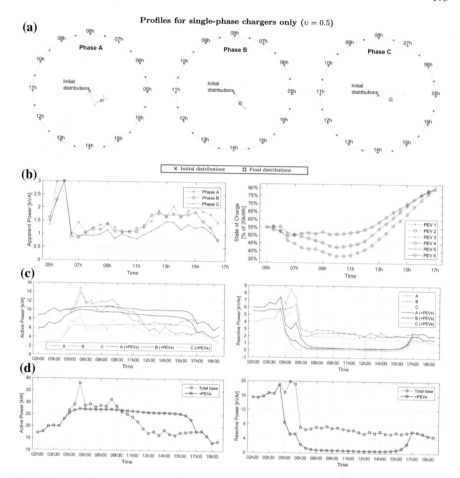

Fig. 5.12 6 PEVs connected to single-phase chargers ($v = 0.5$), by pairs on each phase, and sharing similar conditions. **a** evolution of the 6 normalized (scaled) apparent power distributions for the three phases. **b** Final apparent power distributions per phase, and state of charge profiles. **c** Forecast and final active and reactive power profiles obtained per phase. **d** Forecast and final total active and reactive power profiles

to the grid, as long as resources are evenly shared among PEVs. Unfortunately, the balancing service among phases is not possible either.

Let us take a look on an additional case where the service parameter is fixed at $v = 0.5$. Thus, the objectives importance is fixed half way between individual phase grid objectives, and total grid objectives.

For this case, the evolution of apparent power distributions per PEV can be observed in Fig. 5.12a. It can be noticed that fairness is still achieved in the sense of individual phases, given that PEVs on each phase still reach a common final distribution on their own phase. It is not a global common final distribution for the three phases (as it was for the case with $\upsilon = 0$), but final apparent power profiles get closer than they got for the extreme case with $\upsilon = 1$. This can also be confirmed with the final state of charge profiles per phase, comparing with those of Fig. 5.10b. Besides, comparing with both extreme cases before, in this case both type of profiles are compensated: individual phases, and total system. This can be observed for active and reactive power profiles on Fig. 5.12c, comparing with those of Fig. 5.11c.

This final test, is revised to check alternatives for the utility grid manager when there is presence of only single-phase chargers distributed among the three phases of the transformer. The utility grid manager may propose economic or social incentives to PEV owners to justify the reduction on the global fairness capabilities of the approach, taking into account that local fairness (per phase) is still achieved.

5.8.3 Example 3: Both Three-Phase and Single-Phase Chargers

Let us consider now 12 PEVs sharing similar conditions: connection time at 05h, departure time at 17h, battery capacity of 20 kWh, initial states of charge soc_0^i of 55%, desired soc_d^i of 80%, chargers power limits of ± 3 kW. Time steps are half hours ($K^i = 24$ for all PEVs) and additional parameters are fixed as $\eta = 0.5$ defined by the grid manager, and $\mu^i = 0.8$ for all PEVs. This time, 6 PEVs are connected to three-phase chargers, and the other 6 are connected to single-phase chargers distributed by pairs per phase. Three values are considered for parameter υ to check its effects on the approach. A summary of these assumptions is given on Table 5.3.

Profiles on Figs. 5.13, 5.14, and 5.15, show the most interesting outcomes for this example when $\upsilon = 1$, $\upsilon = 0.5$, and $\upsilon = 0$, respectively. Again in these tests, 50 rounds of PEVs-Aggregator exchanges of information are applied.

Table 5.3 Summary of Example 3: both single and three-phase chargers

Item	Description
Number of vehicles	12 PEVs
Chargers	All of them 3 kW/phase
Batteries	All of them 20 kWh
Initial states of charge	$soc_0^i = 55\%$
Desired states of charge	$soc_d^i = 80\%$
Utility grid manager	Service parameter fixed at $\eta = 0.5$
Trade-off factor	$\mu^i = 0.8$ for all PEVs
Type of chargers	6 Three-phase chargers, and 6 single-phase chargers
Time period	24 half hour steps ($K^i = 24$) between 05 and 17h (disconnection at 17h)

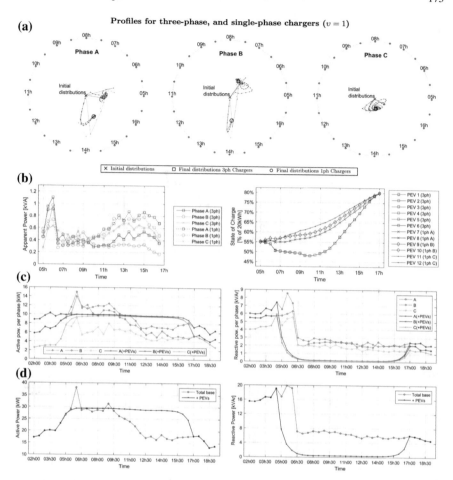

Fig. 5.13 12 PEVs, 6 connected to three-phase charger, and 6 connected to single-phase chargers by pairs per phase. PEVs share similar conditions and parameter $v = 1$. **a** evolution of the 12 normalized (scaled) apparent power distributions per phase. **b** Final apparent power distributions per phase, and state of charge profiles per type of charger and phase. **c** Forecast and final active and reactive power profiles per phase. **d** Forecast and final total active and reactive power profiles

Again, resources and tasks are fairly shared among PEVs per phase, according to the type of charger and the parameter v. In all the cases, three-phase chargers converge to common apparent power distributions on each of their phases, as it can be observed on Figs. 5.13a, 5.14a, and 5.15a. Single-phase chargers also converge to common apparent power distributions per phase. These final apparent power distributions per phase for single-phase chargers change according to the v value. They get closer to each other with the decrease of v, until they become identical for the three phases when $v = 0$. This can be observed as well, on Figs. 5.13a, 5.14a, and 5.15a.

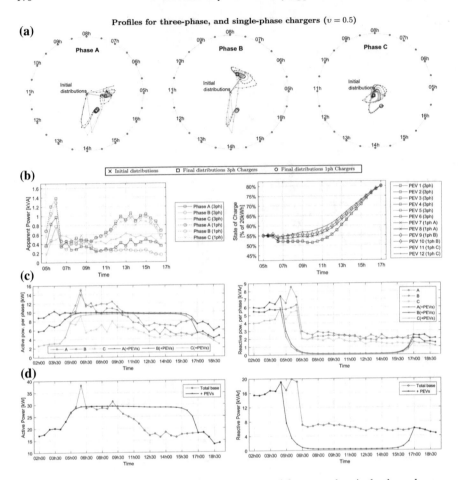

Fig. 5.14 12 PEVs, 6 connected to three-phase charger, and 6 connected to single-phase chargers by pairs per phase. PEVs share similar conditions and parameter $v = 0.5$. **a** evolution of the 12 normalized (scaled) apparent power distributions per phase. **b** Final apparent power distributions per phase, and state of charge profiles per type of charger and phase. **c** Forecast and final active and reactive power profiles per phase. **d** Forecast and final total active and reactive power profiles

All type of chargers start with uniform apparent power distributions per phase, and with the evolution driven by the ED approach, all of them converge to the corresponding common final distribution. The final apparent power distributions, can be checked on Figs. 5.13b, 5.14b, and 5.15b.

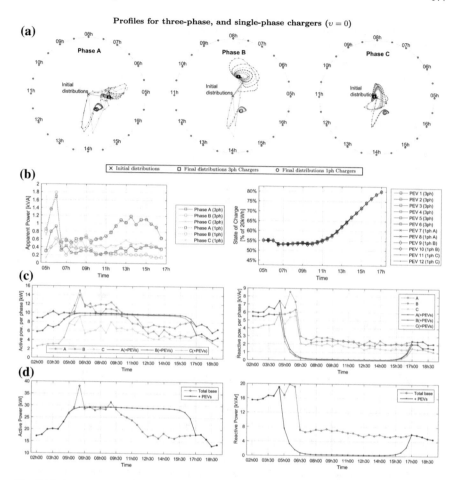

Fig. 5.15 12 PEVs, 6 connected to three-phase charger, and 6 connected to single-phase chargers by pairs per phase. PEVs share similar conditions and parameter $v = 0$. **a** evolution of the 12 normalized (scaled) apparent power distributions per phase. **b** Final apparent power distributions per phase, and state of charge profiles per type of charger and phase. **c** Forecast and final active and reactive power profiles per phase. **d** Forecast and final total active and reactive power profiles

Given that variations on v modify the final common apparent power distributions where single-phase chargers converge, these variations also modify the distributions where three-phase chargers converge. That is because all PEVs interact with each other, no matter the type of charger. The parameter v only isolates single-phase chargers in a phase from single-phase chargers in other phases (if it is $v = 1$). This behavior can be verified directly by comparing the final apparent power profiles for three-phase chargers (on each of their phases) and single-phase chargers on Figs. 5.13b, 5.14b, and 5.15b.

A simple way to check the effect of v over both type of chargers, is to revise the final state of charge profiles for three-phase and single-phase chargers in phases A, B, and C, for each of the considered values $v = \{1, 0.5, 0\}$. For the isolating value $v = 1$, on Fig. 5.13b, it can be noticed that three-phase chargers end up with the same state of charge profile discharging down to 47.9% until 10h, and then charging all the time until the disconnection time, at 17h. Single-phase chargers never reach states of charge below the initial 55%. The pair of PEVs in phase C reach higher states of charge earlier in the connection time (at 11h, chargers in phase C have charged up to 64.4%, while those in phase A and B have reached 59% and 61.1%, respectively).

When $v = 0.5$, on Fig. 5.14b, it can be noticed as well that all three-phase chargers end up with the same state of charge profile. This time they discharge down to 51.4% until 9h30, and then they charge all the time until the disconnection time, at 17h. Single-phase chargers reach states of charge below 55%. The pair of PEVs on phase A reaches the lower state of charge (53.5% at 6h 30). Chargers in phase C are still privileged, reaching higher states of charge earlier than three-phase chargers and single-phase chargers on other phases (up to 59% at 11h, while those in phase A and B have reached 56.1% and 57.3%, respectively).

Finally, with $v = 0$, on Fig. 5.15b, it can be observed that the state of charge profiles for all type of chargers become almost identical. All of them discharge until 6h30 up to 52.9%, and at 13h all of them have reached 62%. The whole time, they share similar charging and discharging rates until they reach the desired 80% at 17h. All the single-phase chargers reach the same common final apparent power distribution, as it can be observed on Fig. 5.15a, b, while three-phase chargers reach common apparent power distributions per phase as well. Even if state of charge profiles for all type of chargers are similar, the final apparent power profiles for single and three-phase chargers deffer because the last type is able to distribute its resources among the three phases rather than just one. It is important to observe that global fairness is achieved among PEVs, not only for distributed resource sharing (state of charge profiles), but for the task of service providing as well. This can be observed in the resulting smooth, flat, and balanced active and reactive power profiles per phase, on Fig. 5.15c, and for the resulting smooth, and flat total active and reactive power profiles on Fig. 5.15d.

On the other hand, comparing the final active and reactive power profiles per phase, and the final total active and reactive power profiles, for the three considered values of the parameter v, it can be observed that the smoothing, flattening, and balancing objectives are achieved almost identically for all the cases (please check Figs. 5.13c–d, 5.14c–d, and 5.15c–d). However, for values of v larger than 0, the sense of *fairness* becomes clustered per phases, depending on the type of charger, and the phase of connection.

5.8.4   Example 4: Under Realistic Conditions

For this final example let us consider historic data measured from a transformer in a *commerce and offices* area, from the SOREA utility grid company [SOR]. The purpose of this example is to test the ED multi-population approach under a realistic scenario. In this case, there is presence of both single and three-phase chargers in equal proportions. Single-phase chargers are uniformly distributed among phases.

Four study days are taken into account, and the random arrival and departure of several PEVs is considered. Arrivals and departures are obtained using a *Poisson* process model with variable rate of arrivals (changing with time) and variable connection times. The number of connected vehicles, the number of vehicles connected to three-phase and single-phase chargers, the number of vehicles connected to single phase-chargers per phase, and the number of arrivals each half hour, can be checked in Fig. 5.16.

On the other hand, the base active and reactive power profiles per phase of the transformer (without PEVs), can be observed in Fig. 5.17. Storage capacities can be 8.8 kWh with a probability of 30%, and 20 kWh with a probability of 70%. Chargers can have a nominal rate of 3.3 kW/phase with a probability of 80%, and 7.4 kW/phase with a probability of 20%. States of charge are limited to be between 25 and 90% for

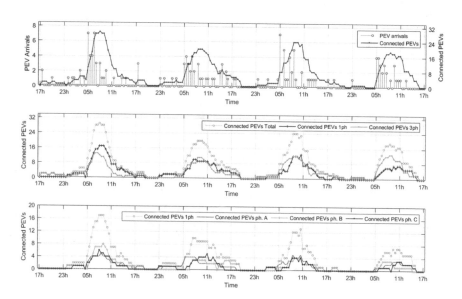

Fig. 5.16 Four days example. Profiles for the total amount of connected vehicles, the amount of vehicles connected to single and three-phase chargers, the amount of vehicles per phase, and the number of arriving PEVs each half hour

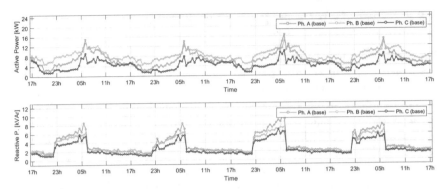

Fig. 5.17 Four days example. Active and reactive power profiles. Forecasts based o real historic data measured from a transformer in a *commerce and offices* area, from the SOREA utility grid company [SOR], France

Table 5.4 Summary and assumptions for Example 4: realistic scenario with single and three-phase chargers

Item	Description
Chargers	3.3 kW/phase with probability of 80%
	7.5 kW/phase with probability of 20%
Batteries	8.8 kWh with probability of 30%
	20 kWh with probability of 70%
Constraints on batteries	Between 25 and 90% for 8.8 kWh
	Between 30 and 85% for 20 kWh
1-ph & 3-phase charger	Probability of 50% for single-phase
	and 50% for three-phase charger
	7.5 kW/phase with probability of 20%
Connection phase	Probability of 33.33% for each phase
Time period	4 days (192 half hour steps)
Highest rate of arrivals	5 PEVs/h at 05h
Lowest rate of arrivals	0.5 PEVs/h at 04 h next day
Peak of connected PEVs	Between 16 and 32 PEVs around 09 h each day
Trade-off values μ^i	Common and fixed for all PEVs to $\mu^i = 0.9$
Utility grid manager	• Service parameter fixed at $\eta = 0.7$
	• Evaluation of multiple values of the service trade-off factor υ
Distribution system info.	Data from SOREA utility grid company [SOR]

8.8 kWh capacities, and between 30 and 85% for the 20 kWh capacities. Parameters are fixed as $\eta = 0.7$ defined by the grid manager, and $\mu^i = 0.9$ for all PEVs, assuming the social or economic incentives for owners (from the utility grid manager) motivate them effectively. A summary of these assumptions is given on Table 5.4.

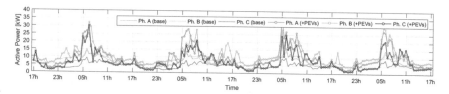

Fig. 5.18 Forecast and final active power profiles per phase, without management. These profiles correspond to the scenario with both three and single-phase chargers

The effect of the service trade-off parameter v is studied here. Two sets of profiles are shown on Figs. 5.19, and 5.20. Each one of these figures corresponds respectively to the definition of the single-phase service trade-off parameter at $v = 1$, $v = 0$. It is possible to compare the final obtained profiles for active power per phase, on Figs. 5.19a, and 5.20a. Beyond some particular small peaks observed when $v = 0$ (due to single-phase chargers), under both scenarios the obtained profiles are very similar. In fact, the presence of three-phase chargers allows to balance the three phases, while at the same time the flattening and smoothing objectives are accomplished.

As it was studied before, when $v = 1$ the individual profiles for each phase are better handled than it is for $v = 0$. However, the final differences are marginal, and in both cases the obtained active power profiles per phase are much more convenient for the transformer and the grid, than without management (see Fig. 5.18).

In terms of reactive powers per phase, the resulting profiles per phase are also better when the efforts of single-phase chargers are focused on individual phases ($v = 1$), as it can be noticed by comparing Figs. 5.19b and 5.20b. However, the differences are again marginal, given the presence of three-phase chargers compensating the effects of single-phase chargers, and balancing the three phases.

On the other hand, checking total active and reactive power profiles, the outcomes are substantially better when $v = 0$ as it can be expected given the efforts of single-phase chargers. However, analyzing Figs. 5.19c–d, and 5.20c–d, the differences are negligible given the presence of three-phase chargers balancing the effects of single-phase chargers on individual phases.

As it has been observed, the presence of three-phase chargers mitigates the effects that the service trade-off parameter v has on the behavior of single-phase chargers, and the resulting profiles. However, as it was concluded with the 12 PEVs example at the beginning of this section, the effect of v is better appreciated in the fair allocation of charging rates among PEVs, with respect to the type of chargers. Comparing state of charge profiles on Figs. 5.19e, and 5.20e, it can be observed that for $v = 0$ profiles are more congruent among each other, no matter the type of charger. These results allow to conclude that if there is a well balanced presence of three-phase and single-

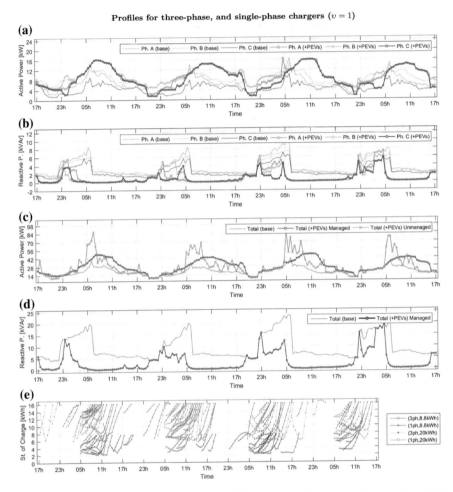

Fig. 5.19 Four days example with $\upsilon = 1$ in the ED approach. **a** Forecast and final active power profiles per phase. **b** Forecast and final reactive power profiles per phase. **c** Forecast and final total active power profiles. Comparison with unmanaged case. **d** Forecast and final total reactive power profiles. **e** State of charge profiles for each PEV

phase chargers, or if three-phase chargers are dominant, then values of υ closer to zero are better for the global fairness of the proposed approach. As it was observed, values of υ closer to the unity, separate the fairness in clusters per phase depending on the type of charger and the phase where they are connected.

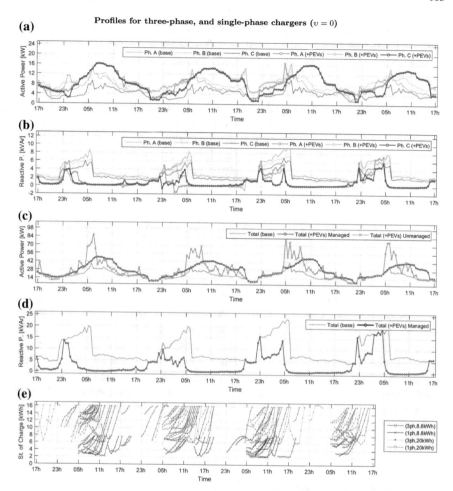

Fig. 5.20 Four days example with $v = 0$ in the ED approach. **a** Forecast and final active power profiles per phase. **b** Forecast and final reactive power profiles per phase. **c** Forecast and final total active power profiles. Comparison with unmanaged case. **d** Forecast and final total reactive power profiles. **e** State of charge profiles for each PEV

5.9 Matlab Scripts

Most of the illustrative example results in this chapter were obtained using fixed step sizes in the discretization of the ED equation. Using fixed step sizes requires forced routine stops when the step results in unfeasible distributions. Consequently, the routines may produce sub-optimal outcomes. However, the scripts provided in this section include a backtracking line-search subroutine, similar to that employed in Chap. 4. This subroutine has been included to adapt the step sizes and increase the routine's convergence speed and accuracy. In this case, the direction of descent employed in the backtracking line-search is provided by the ED gradient. It is up to the reader to use this subroutine or not, and to test the differences in terms of performance.

This section provides Matlab scripts for the local ED routines for energy and reactive power populations, and for both single and three-phase chargers. Section 5.9.1 describes the scripts for the local ED routines for energy populations with single and three-phase chargers. Section 5.9.2 describes the scripts for the local ED routines for reactive power populations in both cases as well. Section 5.9.3 provides a script for simulating the scenario, and the ideal operation of the aggregator, in Examples 1,2, and 3 of Sect. 5.8. Finally, an additional ED routine is provided in Sect. 5.9.4 to represent energy population using logarithmic barrier functions for certain constraints.

5.9.1   Scripts for the Local ED Routine: Energy Populations

The following are the local ED routines that each PEV runs in order to find its optimal energy distribution according to the parameters defined by the owner, and the load forecast provided by the aggregator. As it was mentioned before, a backtracking line-search subroutine, is included to have adaptive step sizes and increase the routine's convergence speed. In this case, the direction of descent employed in the backtracking line-search is provided by the ED gradient. It is up to the reader to use this subroutine or not, and test the differences in terms of performance. To do this, it is possible to eliminate the code section named `%% Subroutine for the backtracking line search`, and replace it with a fixed value for the step size `Ts`. However, in doing so, a portion of code must be including in order to verify that constraints are never violated during the evolution using the chosen fixed step size. Two scripts are provided, one for three-phase and one for single-phase chargers.

5.9.1.1   Script for Three-Phase Chargers

This first script is intended to be used for energy distributions with three-phase chargers. At the end of the script, a portion of code is included (commented) to plot the evolution of the distribution with respect to the base forecast load of each phase.

```
function xhat = Book_ED_P_3ph(soc_d,soc_0,soc_max,soc_min,p_max,l_max,dt,...
                              Ki,Lhat_,x0,mu,eta,init_steps)

% ------ Desciption inputs
% soc_d : final desired state of charge [kWh]
% soc_0 : initial state of charge [kWh]
% soc_max : max allowed state of charge in the connection time frame [kWh]
% soc_min : min allowed state of charge in the connection time frame [kWh]
% p_max : max charging/discharging rate [kW] (per phase)
% l_max : max transformed load allowed [kW]
% dt : time step
% Ki : total time steps
% Lhat_ : Load forecast from (t_0-1) to the available horizon (>Ki+1)
%         (array of 3 rows and Ki+2 columns or more)
```

```
% x0 : previously defined load distribution
%         (array of 3 rows and Ki columns)
% mu : owner's chosen trade-off factor in the interval [0,1]
% eta : grid operator chosen trade-off factor in the interval [0,1]
%         ("profile smoothing" or "load shifting - peak shaving")
% init_steps : (1) first round of agregator -PEV information  exhange
%                 (2) second round of agregator -PEV information  exhange,
%                     useful for reseting initial distribution if
%                     transformed is constrained
%                 (0) normal ED routine

% ------ Description outputs
% xhat : final load distribution for the three phases
%         (array of 3 rows and Ki columns)

% load profile (from aggregator) for the PEV connection time
Lhat = Lhat_(:,2:Ki+1);

% Different from MSD, in the ED routine Gamma and psk are energy
% not active power
Gamma = soc_d-soc_0;

% Threshold for accomodating load when the transformer has reached or will
% reach its limit at a given time slot
threshold = 0.1;

% Arrays of non-changing constraints
% (Sect. 5.6.1, Eqs. 5.15c and d)
p_up1 = dt*ones(3,Ki)*p_max; % nominal from charger
p_up2 = dt*max((l_max-(Lhat-x0)),threshold); % available from transformer
p_lo_ = -dt*ones(3,Ki)*p_max; % lower limit also nominal from charger

% variance tolerance
tol = 1e-10;

% Parameters of backtracking line search
alpha_ls = 0.5;
beta_ls = 0.1;

% Number of internal ED routine iterations
It = 1000;

% Initial vector of upper non-changing constraints
p_up_ = min(p_up1,p_up2);

% Definition of initial population distribution
% This is defined in two steps:
% 1) (init_steps=1) The distribution is initialized as a uniform
%     distribution over the connection time frame, and then it is sent to
%     the aggregator.
% 2) (init_steps=2) Once the aggregator receives all the initial
%     distributions, it returns the new aggregated total load profile. In
%     this second step, the PEV resets its load distribution according to
%     the availability of the transformer during the connection time frame.
if init_steps == 1
    % -- First initialization (first step)
    % distribution is defined using charger's nominal constraints (p_up1)
    psk_ = (p_up1/sum(sum(p_up1)))*(soc_d-soc_0);
    % the distribution is assigned
    psk = psk_;
    % after this, the distribution is sent to the aggregator without further
    % treatment
elseif init_steps == 2
```

```
% -- Second initialization (second step) taking into account possible
% transformer constraints
if sum(sum(p_up_)) > Gamma
    % If constraints from charger and transformer allow it, load is
    % distributed such that the initally required energy for the PEV is
    % consumed during the connection time frame
    psk_ = (p_up_/sum(sum(p_up_)))*(soc_d-soc_0);
else
    % otherwise, constraints indicate that there is a lack of energy
    % available. Thus the PEV consumes what is available in a load
    % distribution given by the imposed constrains
    psk_ = (p_up_/sum(sum(p_up_)))*(sum(sum(p_up_))-1e-3);
    % thus the population size is smaller that the initially defined
    Gamma = sum(sum(p_up_))-1e-3;
end

% the second reset distribution is assigned
psk = psk_;
% after this, the distribution is re-sent to the aggregator without
% further treatment
else
    % After the initial resets, load is again distributed according to
    % constraints as in step 2
    if sum(sum(p_up_)) > Gamma
        % enough available energy
        psk_ = (p_up_/sum(sum(p_up_)))*(soc_d-soc_0);
    else
        % lack of energy available
        psk_ = (p_up_/sum(sum(p_up_)))*(sum(sum(p_up_))-1e-3);
        % thus the population size is smaller that the initially defined
        Gamma = sum(sum(p_up_))-1e-3;
    end
    % load distribution is assigned
    psk = psk_;
    % then, it is processed by the ED routine
end

% construction of arrays of changing constraints
% (Sect. 5.6.1, Eqs. 5.15a and b)
soc_k = repmat(soc_0+cumsum(sum(psk_)),3,1);

tri1 = tril(ones(Ki));
tri1 = [tri1;tri1;tri1];
tri1 = reshape(tri1(:),Ki,3*Ki);
tri2 = triu(ones(Ki),1);
tri2 = [tri2;tri2;tri2];
tri2 = reshape(tri2(:),Ki,3*Ki);

UP = soc_max-repmat(soc_k',1,Ki).*tri1+repmat(psk_(:)',Ki,1).*tri1 ...
    +1e6*tri2;
LO = soc_min-repmat(soc_k',1,Ki).*tri1+repmat(psk_(:)',Ki,1).*tri1 ...
    -1e6*tri2;
LO(end,:) = LO(end,:)-soc_min+soc_d-1;

UP = reshape(min(UP),3,Ki);
LO = reshape(max(LO),3,Ki);

% intersect arrays of non-changing and changing constraints
% (step repeated each iteration)
p_up = min(p_up_,UP);
p_lo = max(p_lo_,LO);

% definition of sigmas (repeated at each iteration) (Sect. 5.4)
```

```
sigma_up = Gamma-sum(sum(p_up));
sigma_lo = Gamma-sum(sum(p_lo));

% escort functions and sum of escort functions (Sect. 5.4)
Escort = (psk_-p_up).*(psk_-p_lo)/(sigma_up*sigma_lo);
SumEscort = sum(sum(Escort));

% payoff functions for pure strategies (Sect. 5.6.2)
bc = (Lhat_(:,1:Ki+2)-[zeros(3,1) x0 zeros(3,1)])';
pi_now = psk_'/dt+bc(2:Ki+1,:);
pi_pre = [zeros(3,1) psk_(:,1:Ki-1)]'/dt+bc(1:Ki,:);
pi_pos = [psk_(:,2:Ki) zeros(3,1)]'/dt+bc(3:Ki+2,:);

fk = -2*eta*mu*(pi_now)-(1-eta)*mu*2*(-pi_pre+2*pi_now-pi_pos)...
     -2*(1-mu)*(psk_'/dt);

% scalar field corresponding to payoff functions (based on Sect. 5.6.2)
F = mu*(eta*sum(sum(pi_now.^2))+(1-eta)*(sum(sum((pi_now-pi_pre).^2,2)...
    +(pi_pos(:,end)-pi_now(:,end)).^2)))+(1-mu)*sum(sum((psk_'/dt).^2));

% weighted average payoff
fmeank = trace(Escort*fk)/SumEscort;

if init_steps == 0
    % if the first two rounds of exchanges (verifying transformer's
    % constraints) have already passed, then the ED routine is run
    for i = 1:It

        % the new distribution is assigned
        psk = psk_;

        %% Subroutine for the backtracking line search
        Ts = 1e5;

        F_ir = Inf;
        % the descent direction (Delta_x) is defined by the ED gradient
        Delta_x = Escort.*(fk'-fmeank);
        if norm(Delta_x(:)) == 0
            % it only occurs when the equilibrium has been perfectly reached
            break
        else
            Delta_x = Delta_x/norm(Delta_x(:)); % normalize the ED gradient
        end

        while (F_ir > (F - alpha_ls*Ts*trace(fk'*Delta_x')))
            psk_ir = psk_ + Ts*Delta_x; % test distribution

            % test state of charge profile
            soc_k_ir = repmat(soc_0+cumsum(sum(psk_ir)),3,1);

            % test arrays of changing constrains
            UP_ir = soc_max-repmat(soc_k_ir',1,Ki).*tri1...
                    +repmat(psk_ir(:)',Ki,1).*tri1+1e6*tri2;
            LO_ir = soc_min-repmat(soc_k_ir',1,Ki).*tri1...
                    +repmat(psk_ir(:)',Ki,1).*tri1-1e6*tri2;
            LO_ir(end,:) = LO_ir(end,:)-soc_min+soc_d-1;
            UP_ir = reshape(min(UP_ir),3,Ki);
            LO_ir = reshape(max(LO_ir),3,Ki);

            % test intersection arrays non-changing and changing constraints
            p_up_ir = min(p_up_,UP_ir);
            p_lo_ir = max(p_lo_,LO_ir);
```

```
    % checks if test distributions lie within specified ranges
    if (sum(sum(psk_ir<p_lo_ir)) + sum(sum(psk_ir>p_up_ir))) > 0
        % it is useless to update the scalar field value if
        % parameters are out of range
    else
        % updates scalar field test value
        pi_now_ir = psk_ir'/dt + bc(2:Ki+1,:);
        pi_pre_ir = [zeros(3,1) psk_ir(:,1:Ki-1)]'/dt + bc(1:Ki,:);
        pi_pos_ir = [psk_ir(:,2:Ki) zeros(3,1)]'/dt + bc(3:Ki+2,:);
        F_ir = mu*(eta*sum(sum(pi_now_ir.^2))...
                +(1-eta)*(sum(sum((pi_now_ir-pi_pre_ir).^2,2)...
                +(pi_pos_ir(:,end)-pi_now_ir(:,end)).^2)))...
                +(1-mu)*sum(sum((psk_ir'/dt).^2));
    end
    % reduces step size Ts
    Ts = beta_ls*Ts;

    % If Ts is too small, the subroutine stops
    if Ts < 1e-30
        break
    end
end
if Ts < 1e- if Ts is too small, the ED routine is stopped
    break
end
%%

% Computes next distribution
psk_ = psk_+Ts*Delta_x;

% Computes next state of charge profile
soc_k = repmat(soc_0+cumsum(sum(psk_)),3,1);

UP = soc_max-repmat(soc_k',1,Ki).*tri1+repmat(psk_(:)',Ki,1).*tri1...
    + 1e6*tri2;
LO = soc_min-repmat(soc_k',1,Ki).*tri1+repmat(psk_(:)',Ki,1).*tri1...
    - 1e6*tri2;
LO(end,:) = LO(end,:)-soc_min+soc_d-1;
UP = reshape(min(UP),3,Ki);
LO = reshape(max(LO),3,Ki);

% computes next intersection of non-changing and changing constraints
p_up = min(p_up_,UP);
p_lo = max(p_lo_,LO);

% computes next sigma values
sigma_up = Gamma-sum(sum(p_up));
sigma_lo = Gamma-sum(sum(p_lo));

% computes next escort function values and their sum
Escort = (psk_-p_up).*(psk_-p_lo)/(sigma_up*sigma_lo);
SumEscort = sum(sum(Escort));

% new payoffs
pi_now = psk_'/dt+bc(2:Ki+1,:);
pi_pre = [zeros(3,1) psk_(:,1:Ki-1)]'/dt+bc(1:Ki,:);
pi_pos = [psk_(:,2:Ki) zeros(3,1)]'/dt+bc(3:Ki+2,:);

fk = -2*eta*mu*(pi_now)-(1-eta)*mu*2*(-pi_pre+2*pi_now-pi_pos)...
    -2*(1-mu)*(psk_'/dt);

% new scalar field value corresponding to payoff functions
F = mu*(eta*sum(sum(pi_now.^2))...
```

```
                        +(1-eta)*(sum(sum((pi_now-pi_pre).^2,2)...
                        +(pi_pos(:,end)-pi_now(:,end)).^2)))...
                        +(1-mu)*sum(sum((psk_'/dt).^2));

            % weighted average payoff
            fmeank = trace(Escort*fk)/SumEscort;

            % updates the weighted variance stopping criteria
            variancef = sum(Escort(:).*(fk(:)...
                            -sum(Escort(:).*fk(:))/SumEscort).^2)/SumEscort;

            % stopping criteria: variance, deviations, or max iterations
            if (SumEscort*variancef < tol) || ((sum(sum(psk_))/Gamma)-1)^2>0.01
                break
            end

            % If you want to check the evolution inside this ED routine
            % please uncomment this sectionof code
%               figure(10)
%               plot(bc,':')
%               hold on
%               plot(bc+[zeros(3,1) psk/dt zeros(3,1)]')
%               hold off
%               grid
%               ylabel('active power [kW]')
%               xlabel('forecast horizon [h]')
%               set(gca,'FontSize',12);
%               title( ['iteration = ' num2str(i) '     variance = '...
%                       num2str(variancef)])
%               pause(1e-6)

        end
    end
    % output assignation
    xhat = psk/dt;
end
```

5.9.1.2   Script for Single-Phase Chargers

This second script is intended to be used for energy distributions with single-phase chargers. At the end of the script, a portion of code is also included (commented) to plot the evolution of the distribution with respect to the base forecast load of the concerned phase where the PEV is connected.

```
function xhat = Book_ED_P_1ph(soc_d,soc_0,soc_max,soc_min,p_max,l_max,dt,...
                        Ki,Lhat_,x0,phase,mu,eta,upsilon,init_steps)

% ------ Desciption inputs
% soc_d : final desired state of charge [kWh]
% soc_0 : initial state of charge [kWh]
% soc_max : max allowed state of charge in the connection time frame [kWh]
% soc_min : min allowed state of charge in the connection time frame [kWh]
% p_max : max charging/discharging rate [kW] (per phase)
% l_max : max transformed load allowed [kW]
% dt : time step
% Ki : total time steps
% Lhat_ : Load forecast from (t_0-1) to the available horizon (>Ki+1)
%               (array of 3 rows and Ki+2 columns or more)
% x0 : previously defined load distribution
```

```
%        (array of 3 rows and Ki columns)
% phase : phase where the PEV is connected (1,2 or 3)
% mu : owner's chosen trade-off factor in the interval [0,1]
% eta : grid operator chosen trade-off factor in the interval [0,1]
%        ("profile smoothing" or "load shifting - peak shaving")
% upsilon : grid operator chosen trade-off factor in the interval [0,1]
%        ("single-phase interests" or "three-phase interests")
% init_steps : (1) first round of agregator -PEV information  exhange
%              (2) second round of agregator -PEV information  exhange,
%                   useful for reseting initial distribution if
%                   transformed is constrained
%              (0) normal ED routine


% ------ Description outputs
% xhat : final load distribution for the three phases
%        (unconcerned phases are filled with zeros)
%        (array of 3 rows and Ki columns)

% load profile (from aggregator) for the PEV connection time
Lhat = Lhat_(:,2:Ki+1);

% Different from MSD, in the ED routine Gamma and psk are energy
% not active power
Gamma = soc_d-soc_0;

% Threshold for accomodating load when the transformer has reached or will
% reach its limit at a given time slot
threshold = 0.1;

% Arrays of non-changing constraints
% (Sect. 5.6.1.1, Eqs. 5.19c and d)
p_up1 = dt*ones(1,Ki)*p_max;
p_up2 = dt*max((l_max*3-sum(Lhat-x0)),threshold);
p_lo_ = -dt*ones(1,Ki)*p_max;

% variance tolerance
tol = 1e-10;

% Parameters of backtracking line search
alpha_ls = 0.5;
beta_ls = 0.1;

% Number of internal ED routine iterations
It = 1000;

% Initial vector of upper non-changing constraints
p_up_ = min(p_up1,p_up2);

% Definition of initial population distribution
% This is defined in two steps:
% 1) (init_steps=1) The distribution is initialized as a uniform
%    distribution over the connection time frame, and then it is sent to
%    the aggregator.
% 2) (init_steps=2) Once the aggregator receives all the initial
%    distributions, it returns the new aggregated total load profile. In
%    this second step, the PEV resets its load distribution according to
%    the availability of the transformer during the connection time frame.
if init_steps == 1
    % -- First initialization (first step)
    % distribution is defined using charger's nominal constraints (p_up1)
    psk_ = (p_up1/sum(p_up1))*(soc_d-soc_0);
    % the distribution is assigned
    psk = psk_;
```

```
        % after this, the distribution is sent to the aggregator without further
        % treatment
elseif init_steps == 2
        % -- Second initialization (second step) taking into account possible
        % transformer constraints
        if sum(p_up_)>Gamma
            % If constraints from charger and transformer allow it, load is
            % distributed such that the initally required energy for the PEV is
            % consumed during the connection time frame
            psk_ = (p_up_/sum(p_up_))*(soc_d-soc_0);
        else
            % otherwise, constraints indicate that there is a lack of energy
            % available. Thus the PEV consumes what is available in a load
            % distribution given by the imposed constrains
            psk_ = (p_up_/sum(p_up_))*(sum(p_up_)-1e-3);
            % thus the population size is smaller that the initially defined
            Gamma = sum(p_up_)-1e-3;
        end

        % the second reset distribution is assigned
        psk = psk_;
        % after this, the distribution is re-sent to the aggregator without
        % further treatment
else
        % After the initial resets, load is again distributed according to
        % constraints as in step 2
        if sum(p_up_)>Gamma
            % enough available energy
            psk_ = (p_up_/sum(p_up_))*(soc_d-soc_0);
        else
            % lack of energy available
            psk_ = (p_up_/sum(p_up_))*(sum(p_up_)-1e-3);
            % thus the population size is smaller that the initially defined
            Gamma = sum(p_up_)-1e-3;
        end
        % load distribution is assigned
        psk = psk_;
        % then, it is processed by the ED routine
end

% construction of arrays of changing constraints
% (Sect. 5.6.1.1, Eqs. 5.19a and b)
soc_k = soc_0+cumsum(psk_);

M1 = tril(ones(Ki))-tril(ones(Ki),-(Ki));
M2 = 1e6*not(tril(ones(Ki))-tril(ones(Ki),-(Ki)));

UP = (soc_max-repmat(soc_k',1,Ki).*M1+repmat(psk_,Ki,1).*M1+M2);
LO = (soc_min-repmat(soc_k',1,Ki).*M1+repmat(psk_,Ki,1).*M1-M2);
LO(end,:) = LO(end,:)-soc_min+soc_d-1;

UP = min(UP(1:Ki,:));
LO = max(LO(1:Ki,:));

% intersect arrays of non-changing and changing constraints
% (step repeated each iteration)
p_up = min(p_up_,UP);
p_lo = max(p_lo_,LO);

% definition of sigmas (repeated at each iteration) (Sect. 5.4)
sigma_up = Gamma-sum(p_up);
sigma_lo = Gamma-sum(p_lo);
```

```
% escort functions and sum of escort functions (Sect. 5.4)
Escort = (psk_-p_up).*(psk_-p_lo)/(sigma_up*sigma_lo);
SumEscort = sum(Escort);

% payoff functions for pure strategies (Sect. 5.6.2.1)
bc = (Lhat_(phase,1:Ki+2)-[0 x0(phase,:) 0])';
bcT = (sum(Lhat_(:,1:Ki+2))-[0 x0(phase,:) 0])';
pi_now = psk_'/dt+bc(2:Ki+1);
pi_pre = [0 psk_(1:Ki-1)]'/dt+bc(1:Ki);
pi_pos = [psk_(2:Ki) 0]'/dt+bc(3:Ki+2);
piT_now = psk_'/dt+bcT(2:Ki+1);
piT_pre = [0 psk_(1:Ki-1)]'/dt+bcT(1:Ki);
piT_pos = [psk_(2:Ki) 0]'/dt+bcT(3:Ki+2);

fk = -2*upsilon*eta*mu*pi_now ...
     -upsilon*(1-eta)*mu*2*(-pi_pre+2*pi_now-pi_pos)...
     -2*(1-upsilon)*eta*mu*(piT_now)...
     -(1-upsilon)*(1-eta)*mu*2*(-piT_pre+2*piT_now-piT_pos)...
     -2*(1-mu)*(psk_'/dt);

% scalar field corresponding to payoff functions (based on Sect. 5.6.2.1)
F = upsilon*mu*(eta*sum(pi_now.^2 )+(1-eta)*(sum((pi_now-pi_pre).^2) ...
    +(pi_pos(end)-pi_now(end)).^2))+(1-upsilon)*mu*(eta*sum(piT_now.^2)...
    +(1-eta)*(sum((piT_now-piT_pre).^2)+(piT_pos(end)-piT_now(end)).^2))...
    +(1-mu)*sum((psk_'/dt).^2);

% weighted average payoff
fmeank = Escort*fk/SumEscort;

if init_steps == 0
    % if the first two rounds of exchanges (verifying transformer's
    % constraints) have already passed, then the ED routine is run
    for i=1:It

        % the new distribution is assigned
        psk = psk_;

        %% Subroutine for the backtracking line search
        Ts = 1e5;

        F_ir = Inf;
        % the descent direction (Delta_x) is defined by the ED gradient
        Delta_x = Escort.*(fk'-fmeank);
        if norm(Delta_x(:)) == 0
            % it only occurs when the equilibrium has been perfectly reached
            break
        else
            Delta_x = Delta_x/norm(Delta_x(:)); % normalize the ED gradient
        end

        while (F_ir > (F - alpha_ls*Ts*(fk'*Delta_x')))
            psk_ir = psk_ + Ts*Delta_x; % test distribution

            % test state of charge profile
            soc_k_ir = soc_0+cumsum(psk_ir);

            % test arrays of changing constrains
            UP_ir = (soc_max-repmat(soc_k_ir',1,Ki).*M1 ...
                    +repmat(psk_ir,Ki,1).*M1+M2);
            LO_ir = (soc_min-repmat(soc_k_ir',1,Ki).*M1 ...
                    +repmat(psk_ir,Ki,1).*M1-M2);
            LO_ir(end,:) = LO_ir(end,:)-soc_min+soc_d-1;
            UP_ir = min(UP_ir(1:Ki,:));
```

```matlab
    LO_ir = max(LO_ir(1:Ki,:));

    % test intersection arrays non-changing and changing constraints
    p_up_ir = min(p_up_,UP_ir);
    p_lo_ir = max(p_lo_,LO_ir);

    % checks if test distributions lie within specified ranges
    if (sum(psk_ir<p_lo_ir) + sum(psk_ir>p_up_ir)) > 0
        % it is useless to update the scalar field value if
        % parameters are out of range
    else
        % updates scalar field test value and step size Ts
        pi_now_ir = psk_ir'/dt+bc(2:Ki+1);
        pi_pre_ir = [0 psk_ir(1:Ki-1)]'/dt+bc(1:Ki);
        pi_pos_ir = [psk_ir(2:Ki) 0]'/dt+bc(3:Ki+2);
        piT_now_ir = psk_ir'/dt+bcT(2:Ki+1);
        piT_pre_ir = [0 psk_ir(1:Ki-1)]'/dt+bcT(1:Ki);
        piT_pos_ir = [psk_ir(2:Ki) 0]'/dt+bcT(3:Ki+2);
        F_ir = upsilon*mu*(eta*sum(pi_now_ir.^2 ) ...
                +(1-eta)*(sum((pi_now_ir-pi_pre_ir).^2) ...
                +(pi_pos_ir(end)-pi_now_ir(end)).^2))...
                +(1-upsilon)*mu*(eta*sum(piT_now_ir.^2) ...
                +(1-eta)*(sum((piT_now_ir-piT_pre_ir).^2) ...
                +(piT_pos_ir(end)-piT_now_ir(end)).^2))...
                +(1-mu)*sum((psk_ir'/dt).^2);
    end
    % reduces step size Ts
    Ts = beta_ls*Ts;

    % If Ts is too small, the subroutine stops
    if Ts < 1e-30
        break
    end
end

if Ts < 1e- if Ts is too small, the ED routine is stopped
    break
end
%%

% Computes next distribution
psk_ = psk_+Ts*Delta_x;

% Computes next state of charge profile
soc_k = soc_0+cumsum(psk_);

UP = (soc_max-repmat(soc_k',1,Ki).*M1+repmat(psk_,Ki,1).*M1+M2);
LO = (soc_min-repmat(soc_k',1,Ki).*M1+repmat(psk_,Ki,1).*M1-M2);
LO(end,:) = LO(end,:)-soc_min+soc_d-1;
UP = min(UP(1:Ki,:));
LO = max(LO(1:Ki,:));

% computes next intersection of non-changing and changing constraints
p_up = min(p_up_,UP);
p_lo = max(p_lo_,LO);

% computes next sigma values
sigma_up=Gamma-sum(p_up);
sigma_lo=Gamma-sum(p_lo);

% computes next escort function values and their sum
Escort=(psk_-p_up).*(psk_-p_lo)/(sigma_up*sigma_lo);
SumEscort=sum(Escort);
```

```
             % new payoffs
             pi_now = psk_'/dt+bc(2:Ki+1);
             pi_pre = [0 psk_(1:Ki-1)]'/dt+bc(1:Ki);
             pi_pos = [psk_(2:Ki) 0]'/dt+bc(3:Ki+2);
             piT_now = psk_'/dt+bcT(2:Ki+1);
             piT_pre = [0 psk_(1:Ki-1)]'/dt+bcT(1:Ki);
             piT_pos = [psk_(2:Ki) 0]'/dt+bcT(3:Ki+2);

             fk = -2*upsilon*eta*mu*(pi_now) ...
                   -upsilon*(1-eta)*mu*2*(-pi_pre+2*pi_now-pi_pos) ...
                   -2*(1-upsilon)*eta*mu*(piT_now) ...
                   -(1-upsilon)*(1-eta)*mu*2*(-piT_pre+2*piT_now-piT_pos) ...
                   -2*(1-mu)*(psk_'/dt);

             % new scalar field value corresponding to payoff functions
             F = upsilon*mu*(eta*sum(pi_now.^2 ) ...
                   +(1-eta)*(sum((pi_now-pi_pre).^2) ...
                   +(pi_pos(end)-pi_now(end)).^2)) ...
                   +(1-upsilon)*mu*(eta*sum(piT_now.^2) ...
                   +(1-eta)*(sum((piT_now-piT_pre).^2) ...
                   +(piT_pos(end)-piT_now(end)).^2)) ...
                   +(1-mu)*sum((psk_'/dt).^2);

             % weighted average payoff
             fmeank=Escort*fk/SumEscort;

             % updates the weighted variance stopping criteria
             variancef = sum(Escort'.*(fk-sum(Escort'.*fk)/SumEscort).^2)/SumEscort;

             % stopping criteria: variance, deviations, or max iterations
             if (SumEscort*variancef<tol)
                  break
             end

             % If you want to check the evolution inside this ED routine
             % please uncomment this sectionof code
%            figure(10)
%            plot(Lhat_(:,1:Ki+2)',':')
%            hold on
%            plot(Lhat_(phase,1:Ki+2)'+[0 psk/dt 0]')
%            hold off
%            grid
%            ylabel('active power [kW]')
%            xlabel('forecast horizon [h]')
%            set(gca,'FontSize',12);
%            title( ['iteration = ' num2str(i) '    variance = '...
%                    num2str(variancef)])
%            pause(1e-6)

      end
end
% output assignation
xout3ph=x0*0;
xout3ph(phase,:)=psk;
xhat=xout3ph/dt;
end
```

5.9.2 Scripts for the Local ED Routine: Reactive Power Populations

The following are the local ED routines that each PEV runs in order to find its optimal reactive power distribution according to the parameters defined by the owner, the load forecast provided by the aggregator, and the previously defined energy distributions. As it was mentioned before, a backtracking line-search subroutine, is included to have adaptive step sizes and increase the routine's convergence speed. Two scripts are provided, one for three-phase and one for single-phase chargers.

5.9.2.1 Script for Three-Phase Chargers

This third script is intended to be used for reactive power distributions with three-phase chargers. At the end of the script, a portion of code is included (commented) to plot the evolution of the distribution with respect to the base forecast of reactive power load per phase.

```
function qhat = Book_ED_Q_3ph(p_max,Ki,Qhat_,x_,y0_,mu,eta,init_steps)

% ------ Desciption inputs
% p_max : max charging/discharging rate [kW] (per phase)
% Ki : total time steps
% Qhat_ : Reactive power forecast from (t_0-1) to a horizon (>Ki+1)
%          (array of 3 rows and Ki+2 columns or more)
% x_ : previously defined energy distribution (given by "Book_EDQ_3ph(...)")
%      (array of 3 rows and Ki columns)
% y0 : previously defined reactive power distribution
%      (array of 3 rows and Ki columns)
% mu : owner's chosen trade-off factor in the interval [0,1]
% eta : grid operator chosen trade-off factor in the interval [0,1]
%       ("profile smoothing" or "load shifting - peak shaving")
% init_steps : (1) first round of agregator -PEV information  exhange
%              (2) second round of agregator -PEV information  exhange,
%                  useful for resetting initial distribution if
%                  transformed is constrained
%              (0) normal ED routine

% ------ Description outputs
% qhat : final reactive power distribution for the three phases
%        (array of 3 rows and Ki columns)

% reactive power profile (from aggregator) for the PEV connection time
% from one time step before to one time step after
Qhat_aux = Qhat_(:,1:Ki+2);
Qhat_aux = Qhat_aux(:)';

% reactive power profile (from aggregator) for the PEV connection time
Qhat = Qhat_(:,2:Ki+1);
Qhat = Qhat(:)';

% express the three arrays (one per phase) as a single row
x = x_(:)';
y0 = y0_(:)';
```

```
% create the vectors of constraints
% (Sect. 5.6.1, Eqs. 5.17a and b)
q_up = ((p_max)^2-x.^2).^0.5;
Q = sum(q_up);
q_up = [q_up Q];
q_lo = -q_up;

% variance tolerance
tol = 1e-10;

% Parameters of backtracking line search
alpha_ls = 0.5;
beta_ls = 0.1;

% Number of internal ED routine iterations
It=1000;

% Definition of initial population distribution
% (independent from the steps that define intial distributions on
% "Book_ED_P_3ph")
% The inital distribution is always zero for each phase and time slot
qsk_ = [Qhat*0 0];
qsk = qsk_;

% Definition of sigma values (Sect. 5.4)
sigma_up=0-sum(q_up);
sigma_lo=0-sum(q_lo);

% escort functions and sum of escort functions (Sect. 5.4)
Escort=(qsk_-q_up).*(qsk_-q_lo)/(sigma_up*sigma_lo);
SumEscort=sum(Escort);

% payoff functions for pure strategies (Sect. 5.6.2)
bc = (Qhat_aux-[zeros(1,3) y0 zeros(1,3)])';
rho_now = qsk_(1:end-1)'+bc(4:end-3);
rho_pre = [zeros(1,3) qsk_(1:end-4)]'+bc(1:end-6);
rho_pos = [qsk_(4:end-1) zeros(1,3)]'+bc(7:end);

hk = -2*eta*mu*rho_now ...
     -2*(1-eta)*mu*(-rho_pre+2*rho_now-rho_pos)-2*(1-mu)*qsk_(1:end-1)';
hk = [hk ; 0];
% scalar field corresponding to payoff functions (based on Sect. 5.6.2)
H = mu*(eta*sum(rho_now.^2)+(1-eta)*(sum((rho_now-rho_pre).^2) ...
    +sum((rho_pos(end-3:end)-rho_now(end-3:end)).^2))) ...
    +(1-mu)*sum(qsk_(1:end-1).^2);

% weighted average payoff
hmeank=Escort*hk/SumEscort;

if init_steps == 0
    % if the first two rounds of exchanges (verifying transformer's
    % constraints) have already passed, then the ED routine is run
    for i=1:It

        % the new distribution is assigned
        qsk = qsk_;

        %% Subroutine for the backtracking line search
        Ts = 1e5;

        H_ir = Inf;
        % the descent direction (Delta_x) is defined by the ED gradient
        Delta_x = Escort.*(hk'-hmeank);
```

```
if norm(Delta_x(:)) == 0
    % it only occurs when the equilibrium has been perfectly reached
    break
else
    Delta_x = Delta_x/norm(Delta_x(:)); % normalize the ED gradient
end

while (H_ir > (H-alpha_ls*Ts*(hk'*Delta_x')))
    qsk_ir = qsk_ + Ts*Delta_x; % test distribution

    % checks if test distributions lie within specified ranges
    if (sum(qsk_ir<q_lo) + sum(qsk_ir>q_up)) > 0
        % it is useless to update the scalar field value if
        % parameters are out of range
    else
        % updates scalar field test value
        rho_now_ir = qsk_ir(1:end-1)'+bc(4:end-3);
        rho_pre_ir = [zeros(1,3) qsk_ir(1:end-4)]'+bc(1:end-6);
        rho_pos_ir = [qsk_ir(4:end-1) zeros(1,3)]'+bc(7:end);
        H_ir = mu*(eta*sum(rho_now_ir.^2) ...
            +(1-eta)*(sum((rho_now_ir-rho_pre_ir).^2) ...
            +sum((rho_pos_ir(end-3:end)-rho_now_ir(end-3:end)).^2)))...
            +(1-mu)*sum(qsk_ir(1:end-1).^2);
    end
    % reduces step size Ts
    Ts = beta_ls*Ts;

    % If Ts is too small, the subroutine stops
    if Ts < 1e-30
        break
if Ts < 1e- if Ts is too small, the ED routine is stopped
    break
end
%%

% Computes next distribution
qsk_ = qsk_+Ts*Delta_x;

% computes next escort function values and their sum
Escort = (qsk_-q_up).*(qsk_-q_lo)/(sigma_up*sigma_lo);
SumEscort = sum(Escort);

% new payoffs
rho_now = qsk_(1:end-1)'+bc(4:end-3);
rho_pre = [zeros(1,3) qsk_(1:end-4)]'+bc(1:end-6);
rho_pos = [qsk_(4:end-1) zeros(1,3)]'+bc(7:end);

hk(1:end-1) = -2*mu*eta*rho_now ...
            -2*(1-eta)*mu*(-rho_pre+2*rho_now-rho_pos)...
            -2*(1-mu)*qsk_(1:end-1)';

% new scalar field value corresponding to payoff functions
H = mu*(eta*sum(rho_now.^2)+(1-eta)*(sum((rho_now-rho_pre).^2) ...
    +sum((rho_pos(end-3:end)-rho_now(end-3:end)).^2)))...
    +(1-mu)*sum(qsk_(1:end-1).^2);

% weighted average payoff
hmeank=Escort*hk/SumEscort;

% updates the weighted variance stopping criteria
varianceh = sum(Escort'.*(hk...
            -sum(Escort'.*hk)/SumEscort).^2)/SumEscort;
```

```
        % stopping criteria: variance, deviations, or max iterations
        if (SumEscort*varianceh<tol)||(abs(sum(qsk_))>0.01)
            break
        end

        % If you want to check the evolution inside this ED routine
        % please uncomment this sectionof code
%           figure(10)
%           plot(reshape(bc',3,Ki+2)',':')
%           hold on
%           plot(reshape(bc',3,Ki+2)'+ ...
%               [zeros(3,1) reshape(qsk(1:end-1),3,Ki) zeros(3,1)]')
%           hold off
%           grid
%           ylabel('reactive power [kVAr]')
%           xlabel('forecast horizon [h]')
%           set(gca,'FontSize',12);
%           title( ['iteration = ' num2str(i) '    variance = '...
%               num2str(varianceh)])
%           pause(1e-6)

    end
end
% output assignation
qhat = reshape(qsk(1:end-1),3,Ki);
end
```

5.9.2.2   Script for Single-Phase Chargers

This fourth script is intended to be used for reactive power distributions with single-phase chargers. again, at the end of the script, a portion of code is included (commented) to plot the evolution of the distribution with respect to the base forecast reactive power load of the concerned phase where the PEV is connected.

```
function qhat = Book_ED_Q_1ph(p_max,Ki,Qhat_,x,y0,phase,mu,eta,upsilon, ...
                        init_steps)

% ------ Desciption inputs
% p_max : max charging/discharging rate [kW] (per phase)
% Ki : total time steps
% Qhat_ : Reactive power forecast from (t_0-1) to a horizon (>Ki+1)
%          (array of 3 rows and Ki+2 columns or more)
% x : previously defined energy distribution (given by "Book_EDQ_1ph(...)")
%      (array of 3 rows and Ki columns)
% y0 : previously defined reactive power distribution
%      (array of 3 rows and Ki columns)
% phase : phase where the PEV is connected (1,2 or 3)
% mu : owner's chosen trade-off factor in the interval [0,1]
% eta : grid operator chosen trade-off factor in the interval [0,1]
%       ("profile smoothing" or "load shifting - peak shaving")
% upsilon : grid operator chosen trade-off factor in the interval [0,1]
%       ("single-phase interests" or "three-phase interests")
% init_steps : (1) first round of agregator -PEV information  exhange
%              (2) second round of agregator -PEV information  exhange,
%                  useful for reseting initial distribution if
%                  transformed is constrained
%              (0) normal ED routine

% ------ Description outputs
```

```matlab
% qhat : final reactive power distribution for the three phases
%         (unconcerned phases are filled with zeros)
%         (array of 3 rows and Ki columns)

% reactive power profile (from aggregator) for the PEV connection time
Qhat = Qhat_(phase,2:Ki+1);

% create the vectors of constraints
% (Sect. 5.6.1.1, Eqs. 5.21a and b)
q_up = ((p_max)^2-x(phase,:).^2).^0.5;
Q = sum(q_up);
q_up = [q_up Q];
q_lo = -q_up;

% variance tolerance
tol = 1e-10;

% Parameters of backtracking line search
alpha_ls = 0.5;
beta_ls = 0.1;

% Number of internal ED routine iterations
It = 1000;

% Definition of initial population distribution
% (independent from the steps that define intial distributions on
% "Book_ED_P_1ph")
% The inital distribution is always zero for each phase and time slot
qsk_ = [Qhat*0 0];
qsk = qsk_;

% Definition of sigma values (Sect. 5.4)
sigma_up = 0-sum(q_up);
sigma_lo = 0-sum(q_lo);

% escort functions and sum of escort functions (Sect. 5.4)
Escort = (qsk_-q_up).*(qsk_-q_lo)/(sigma_up*sigma_lo);
SumEscort = sum(Escort);

% payoff functions for pure strategies (Sect. 5.6.2.1)
bc = (Qhat_(phase,1:Ki+2)-[0 y0(phase,:) 0])';
bcT = (sum(Qhat_(:,1:Ki+2))-[0 y0(phase,:) 0])';
rho_now = qsk_(1:end-1)'+bc(2:end-1);
rho_pre = [0 qsk_(1:end-2)]'+bc(1:end-2);
rho_pos = [qsk_(2:end-1) 0]'+bc(3:end);
rhoT_now = qsk_(1:end-1)'+bcT(2:end-1);
rhoT_pre = [0 qsk_(1:end-2)]'+bcT(1:end-2);
rhoT_pos = [qsk_(2:end-1) 0]'+bcT(3:end);

hk = -2*upsilon*eta*mu*rho_now ...
    -upsilon*(1-eta)*mu*2*(-rho_pre+2*rho_now-rho_pos) ...
    -2*(1-upsilon)*eta*mu*rhoT_now ...
    -(1-upsilon)*(1-eta)*mu*2*(-rhoT_pre+2*rhoT_now-rhoT_pos)...
    -2*(1-mu)*qsk_(1:end-1)';

hk = [hk;0];

% scalar field corresponding to payoff functions (based on Sect. 5.6.2.1)
H = upsilon*mu*(eta*sum(rho_now.^2 )+(1-eta)*(sum((rho_now-rho_pre).^2)...
    +(rho_pos(end)-rho_now(end)).^2)) ...
    +(1-upsilon)*mu*(eta*sum(rhoT_now.^2) ...
    +(1-eta)*(sum((rhoT_now-rhoT_pre).^2) ...
    +(rhoT_pos(end)-rhoT_now(end)).^2)) ...
```

```
        +(1-mu)*sum((qsk_(1:end-1)').^2);

% weighted average payoff
hmeank=Escort*hk/SumEscort;

if init_steps == 0
    % if the first two rounds of exchanges (verifying transformer's
    % constraints) have already passed, then the ED routine is run
    for i=1:It

        % the new distribution is assigned
        qsk=qsk_;

        %% Subroutine for the backtracking line search
        Ts = 1e5;

        H_ir = Inf;
        % the descent direction (Delta_x) is defined by the ED gradient
        Delta_x = Escort.*(hk'-hmeank);

        if norm(Delta_x(:)) == 0
            % it only occurs when the equilibrium has been perfectly reached
            break
        else
            Delta_x = Delta_x/norm(Delta_x(:)); % normalize the ED gradient
        end

        while (H_ir > (H-alpha_ls*Ts*(hk'*Delta_x')))
            qsk_ir = qsk_ + Ts*Delta_x;

            % checks if test distributions lie within specified ranges
            if (sum(qsk_ir<q_lo) + sum(qsk_ir>q_up)) > 0
                % it is useless to update the scalar field value if
                % parameters are out of range
            else
                % updates scalar field test value
                rho_now_ir = qsk_ir(1:end-1)'+bc(2:end-1);
                rho_pre_ir = [0 qsk_ir(1:end-2)]'+bc(1:end-2);
                rho_pos_ir = [qsk_ir(2:end-1) 0]'+bc(3:end);
                rhoT_now_ir = qsk_ir(1:end-1)'+bcT(2:end-1);
                rhoT_pre_ir = [0 qsk_ir(1:end-2)]'+bcT(1:end-2);
                rhoT_pos_ir = [qsk_ir(2:end-1) 0]'+bcT(3:end);

                H_ir = upsilon*mu*(eta*sum(rho_now_ir.^2) ...

                    +(1-eta)*(sum((rho_now_ir-rho_pre_ir).^2) ...
                    +(rho_pos_ir(end)-rho_now_ir(end)).^2)) ...
                    +(1-upsilon)*mu*(eta*sum(rhoT_now_ir.^2) ...
                    +(1-eta)*(sum((rhoT_now_ir-rhoT_pre_ir).^2) ...
                    +(rhoT_pos_ir(end)-rhoT_now_ir(end)).^2)) ...
                    +(1-mu)*sum((qsk_ir(1:end-1)').^2);
            end
            % reduces step size Ts
            Ts = beta_ls*Ts;

            % If Ts is too small, the subroutine stops
            if Ts < 1e-30
                break
            end
        end
        if Ts < 1e- if Ts is too small, the ED routine is stopped
            break
        end
```

```
%%

% Computes next distribution
qsk_=qsk_+Ts*Delta_x;

% computes next escort function values and their sum
Escort=(qsk_-q_up).*(qsk_-q_lo)/(sigma_up*sigma_lo);
SumEscort=sum(Escort);

% new payoffs
rho_now = qsk_(1:end-1)'+bc(2:end-1);
rho_pre = [0 qsk_(1:end-2)]'+bc(1:end-2);
rho_pos = [qsk_(2:end-1) 0]'+bc(3:end);
rhoT_now = qsk_(1:end-1)'+bcT(2:end-1);
rhoT_pre = [0 qsk_(1:end-2)]'+bcT(1:end-2);
rhoT_pos = [qsk_(2:end-1) 0]'+bcT(3:end);
hk(1:end-1) = -2*upsilon*eta*mu*rho_now ...
        -upsilon*(1-eta)*mu*2*(-rho_pre+2*rho_now-rho_pos)...
        -2*(1-upsilon)*eta*mu*(rhoT_now) ...
        -(1-upsilon)*(1-eta)*mu*2*(-rhoT_pre+2*rhoT_now-rhoT_pos) ...
        -2*(1-mu)*qsk_(1:end-1)';

% new scalar field value corresponding to payoff functions
H = upsilon*mu*(eta*sum(rho_now.^2 ) ...
        +(1-eta)*(sum((rho_now-rho_pre).^2) ...
        +(rho_pos(end)-rho_now(end)).^2)) ...
        +(1-upsilon)*mu*(eta*sum(rhoT_now.^2) ...
        +(1-eta)*(sum((rhoT_now-rhoT_pre).^2) ...
        +(rhoT_pos(end)-rhoT_now(end)).^2)) ...
        +(1-mu)*sum((qsk_(1:end-1)').^2);

% weighted average payoff
hmeank=Escort*hk/SumEscort;

% updates the weighted variance stopping criteria
varianceh = sum(Escort'.*(hk ...
            -sum(Escort'.*hk)/SumEscort).^2)/SumEscort;

% stopping criteria: variance, deviations, or max iterations
if (SumEscort*varianceh<tol)||(abs(sum(qsk_))>0.01)
    break
end

% If you want to check the evolution inside this ED routine
% please uncomment this sectionof code
%       figure(10)
%       plot(Qhat_(:,1:Ki+2)',':')
%       hold on
%       plot(Qhat_(phase,1:Ki+2)'+[0 qsk(1:end-1) 0]')
%       hold off
%       grid
%       ylabel('reactive power [kVAr]')
%       xlabel('forecast horizon [h]')
%       set(gca,'FontSize',12);
%       title( ['iteration = ' num2str(i) '    variance = '...
%               num2str(varianceh)])
%       pause(1e-6)

    end
end
% output assignation
yout3ph = y0*0;
yout3ph(phase,:) = qsk(1:end-1);
```

```
qhat = yout3ph;
end
```

5.9.3   Script for the Simulation Scenario

The following is the data employed in Examples 1,2 and 3 of Sect. 5.8. We encourage
the reader to play with the inputs of the example to validate or improve the provided
scripts.

```
close all
clear
clc

%%% Inputs

% Total number of PEVs
Num_PEVs = 6; % for this example it may 6 PEVs or 12PEVs
                % The reader is encouraged to use this template to create
                % different scenarios
ph_case = 3; % (1): single-phase chargers (if Num_PEVs = 6)
                % (3): three-phase chargers (if Num_PEVs = 6)
                % (otherwise): three-phase chargers (if Num_PEVs = 6)

% - If Num_PEVs = 6, then variable ph_case defines if the 6 PEVs have
%    single or three-phase chargers.
% - If Num_PEVs = 12, then variable ph_case is ignored
% - If Num_PEVs has other value, then the 6PEVs/3ph case scenario is chosen
if Num_PEVs == 12
    % num of phases of the charger
    phases =[3 3 3 3 3 3 1 1 1 1 1 1]';
    % phase where a single-phase charger is connected
    phase =[1 1 1 1 1 1 1 2 2 3 3]';
elseif Num_PEVs == 6
    if ph_case == 1
        phases = [1 1 1 1 1 1]';
        phase = [1 1 2 2 3 3]';
    elseif ph_case == 3
        phases=[3 3 3 3 3 3];
    else
        phases=[3 3 3 3 3 3];
    end
else
    J = 6;
    phases=[3 3 3 3 3 3];
end

% base forecast active and reactive power profiles per phase
l_3ph_profile = ...
    [8.526 9.178 11.618 14.972 11.394 12.076 10.198 9.628 10.9 10.904 10.554 ...
    9.336 8.98 7.716 7.628 6.76 5.518 6.036 4.988 6.39 5.424 4.904 5.212 ...
    5.008 4.672 5.268 5.534 4.644 4.046 4.602 4.27 2.304 2.51 2.21 2.514 ...
    2.658 3.492 4.02 4.358 3.476 4.352 3.736 4.162 4.922 4.512 5.332 5.562 ...
    6.234 7.048;
    10.008 9.868 10.26 13.962 11.75 11.268 11.614 11.906 11.824 12.41 11.268 ...
    10.13 9.772 9.692 8.636 6.992 6.954 7.466 7.1 7.076 7 6.752 7.67 7.814 ...
    8.236 7.258 7.692 7.11 5.758 6.476 6.672 4.644 5.166 4.982 5.02 5.042 ...
    6.166 6.872 6.934 6.306 6.89 6.852 7.464 7.606 7.396 7.972 8.036 8.394 ...
```

```
    8.27;
    4.468 7.578 7.258 9.156 5.466 5.906 6.174 7.16 5.62 7.988 6.59 6.682 ...
    6.276 6.474 5.69 5.344 4.624 4.26 4.272 4.94 4.412 4.688 4.524 4.79 ...
    4.886 5.438 4.906 4.384 3.492 2.782 2.532 1.616 1.478 1.57 1.474 1.426 ...
    2.472 2.382 2.55 1.828 1.552 1.784 1.85 2.034 2.196 2.378 2.332 2.656 ...
    3.632];

q_3ph_profile = ...
    [7.414 6.542 8.552 7.312 2.988 2.67 2.522 2.922 2.784 2.842 2.68 2.456 ...
    2.498 2.304 2.296 2.242 1.926 1.9 1.818 2.502 2.038 2.008 2.082 2.018 ...
    1.982 2.06 2.054 1.578 1.614 1.494 1.898 1.452 1.402 1.502 1.548 1.37 ...
    2.908 2.988 4.876 3.762 4.136 3.886 4.504 5.01 5.362 5.212 5.466 6.422 ...
    7.05;
    5.726 5.756 5.956 6.308 2.07 1.932 2.204 2.354 2.35 2.462 2.416 2.466 ...
    2.384 2.404 2.418 1.868 2.092 2.474 2.2 1.966 2 2.004 2.152 2.016 2.27 ...
    2.438 2.372 2.448 2.022 2.072 2.144 1.506 1.656 1.67 1.626 1.704 2.954 ...
    3.424 4.436 3.818 3.884 3.864 4.444 4.62 4.862 4.92 5.194 5.658 5.7;
    5.912 4.526 5.36 5.54 2.046 1.922 2.02 1.928 1.826 2.086 1.764 2.132 ...
    1.788 1.662 1.622 1.712 1.346 1.602 1.556 1.46 1.38 1.336 1.452 1.39 ...
    1.38 1.548 1.394 1.426 1.258 1.086 1.204 1.008 1.052 1.186 1.06 1.006 ...
    2.166 2.268 3.408 2.518 2.556 2.576 3.232 3.358 3.588 3.664 3.724 4.496 ...
    5.588];

% Time steps of arrival for each PEV
t_arr = ones(Num_PEVs,1)*2;

% Time steps of departure for each PEVs in the considered days
t_dep = ones(Num_PEVs,1)*25;

% initial, desired , min, and max states of charge correspond to the absolute
% available capacity (-) of 20KWh
% (i.e., the available capacity per PEV is 11kWh)

% Initial states of charge for each PEV
soc_0 = ones(Num_PEVs,1)*5; % All of them start with  55% (11kWh)

% Desired states of charge for each PEV
soc_d = ones(Num_PEVs,1)*10; % all of them have 20kWh batteries and they seek
                             % to reach 80% (16kWh)

% max state of charge for each PEV
soc_max = ones(Num_PEVs,1)*11; % max is  85% of 20kWh (17kWh)

% min state of charge for each PEV
soc_min = ones(Num_PEVs,1)*0; % min is  30% of 20kWh (6kWh)

% Nominal rates of charge for each PEV
p_max = ones(Num_PEVs,1)*3; % kW per phase

% max load for the transformer
l_max = 100; % kW per phase

% summary of arrival and departure times
t_temps = [t_arr t_dep];

mu_avg = 0.8; % it is assumed that the trade-off factor is common
              % to all PEVs. However the reader is encouraged to
              % change this common value or randomize ir following
              % any rule
eta = 0.5; % defined by the grid operator as described in Sect. 5.7

upsilon = 1; % defined by the grid operator as described in Sect. 5.7
```

```
% Total number of steps on the load profiles
T = length(l_3ph_profile);

% Time step duration (in hours or fractions)
dt = 0.5; % 30mins

% Number of steps in a day
T_day = 48;

% Initial hour in the intial day (5am in this case)
h_0 = 05; %hours (values from 0\,h to 23h)

% Arrays where PEV load is individually placed by the aggregator
xhat_a = zeros(Num_PEVs,T);
xhat_b = zeros(Num_PEVs,T);
xhat_c = zeros(Num_PEVs,T);

% Arrays where PEV reactive power is individually placed by the aggregator
qhat_a = zeros(Num_PEVs,T);
qhat_b = zeros(Num_PEVs,T);
qhat_c = zeros(Num_PEVs,T);
% auxiliary variable for identifying the connected PEVs
index = 0;
% Array for storing the evolution of the initial state of charge which
% increases with the evolution of time
soc_0D = soc_0;

% Auxiliary variable representing the local memory of each PEV storing its
% current optimal distributions
xy_mem_l = cell(Num_PEVs,1);
xy_mem_q = cell(Num_PEVs,1);

% Number of exchanges of information between the aggregator and the and the
% connected PEVs per each time step
Num_exchanges = 50;

% Loop simulating the evolution of time and the interactions between
% Aggregator and PEVs

% variable indicating rounds of initialization for distributions
% Definition of initial population distribution
% Defined in two rounds:
% 1) (init_steps=1) each PEV initializes its energy distribution as a uniform
%    distribution over the connection time frame, and then it is sent to
%    the aggregator.
% 2) (init_steps=2) Once the aggregator receives all the initial
%    distributions, it returns the new aggregated total load profile. In
%    this second step, PEVs reset their energy distribution according to
%    the availability of the transformer during the connection time frame.
init_steps = 0;

for t = 2% only one time step is considered in this example.
            % We encourage the reader to create new scenarios from this
            % template and those from previous chapters

    % connected vehicles
    index = find(t>=t_temps(:,1) & t<=t_temps(:,2));
    if isempty(index)
        % no PEVs connected
    else
        for exchange = 1:Num_exchanges % Num_exchanges of info per time step
            for j = 1:length(index)
                if exchange == 1% the start of a new round of exchanges
```

```
          % PEVs set their initial distributions to check
          % transformer availability
          init_steps = 1;
elseif exchange == 2% in the second exhange PEVs update
          % their inital distributions to match with transformer's
          % constraints
          init_steps = 2;
else
          init_steps = 0;
end
% the number ID associated to the current PEV
m = index(j);
% the aggregatr aggregates the current PEV load profiles to
% the forecast and send it to the current PEV
aggr_l_prof = l_3ph_profile ...
                    +[sum(xhat_a);sum(xhat_b);sum(xhat_c)];
aggr_q_prof = q_3ph_profile ...
                    +[sum(qhat_a);sum(qhat_b);sum(qhat_c)];

% the aggregator verifies if the PEV is able to
% redistribute its load or if it is already suing its time
% slots left to the fullest (p_max)
%
% Variable (slots +1) is the amount of time slots required
% to reach the desired state of charge
slots = ...
 max(floor((soc_d(m)-soc_0D(m))/(dt*phases(m)*p_max(m))),0);
if (t_dep(m)-t+1>=slots+1 && (t_dep(m)-t+1)>1)
     % if the PEV is able to redistribute, then the
     % aggregator sends the information. Then the PEV runs
     % its local ED routines :
     if phases(m)== for 3ph chargers
         xhat_aux = Book_ED_P_3ph(soc_d(m),soc_0D(m),...
                          soc_max(m),soc_min(m),p_max(m),...
                          l_max,dt,t_dep(m)-t+1,...
                          aggr_l_prof(:,t-1:t_dep(m)+1),...
                          [xhat_a(m,t:t_dep(m));...
                          xhat_b(m,t:t_dep(m));...
                          xhat_c(m,t:t_dep(m))],mu_avg,eta,...
                          init_steps);
         qhat_aux = Book_ED_Q_3ph(p_max(m),t_dep(m)-t+1,...
                          aggr_q_prof(:,t-1:t_dep(m)+1),...
                          xhat_aux,[qhat_a(m,t:t_dep(m));...
                          qhat_b(m,t:t_dep(m));...
                          qhat_c(m,t:t_dep(m))],mu_avg,eta,...
                          init_steps);
     else  % for 1ph chargers
         xhat_aux = Book_ED_P_1ph(soc_d(m),soc_0D(m),...
                          soc_max(m),soc_min(m),p_max(m),l_max,...
                          dt,t_dep(m)-t+1,...
                          aggr_l_prof(:,t-1:t_dep(m)+1),...
                          [xhat_a(m,t:t_dep(m));...
                          xhat_b(m,t:t_dep(m));...
                          xhat_c(m,t:t_dep(m))],phase(m),mu_avg,...
                          eta,upsilon,init_steps);
         qhat_aux = Book_ED_Q_1ph(p_max(m),t_dep(m)-t+1,...
                          aggr_q_prof(:,t-1:t_dep(m)+1),...
                          xhat_aux,[qhat_a(m,t:t_dep(m));...
                          qhat_b(m,t:t_dep(m));...
                          qhat_c(m,t:t_dep(m))],phase(m),mu_avg,...
                          eta,upsilon,init_steps);
     end
        % Outputs are assigned
```

```
                     xhat_a(m,t:t_dep(m)) = xhat_aux(1,:);
                     xhat_b(m,t:t_dep(m)) = xhat_aux(2,:);
                     xhat_c(m,t:t_dep(m)) = xhat_aux(3,:);

                     qhat_a(m,t:t_dep(m)) = qhat_aux(1,:);
                     qhat_b(m,t:t_dep(m)) = qhat_aux(2,:);
                     qhat_c(m,t:t_dep(m)) = qhat_aux(3,:);

                     xy_mem_l{j,exchange} = xhat_aux;
                     xy_mem_q{j,exchange} = qhat_aux;
            else
            end
    end
% plot base forecasts and aggregated PEV load (active and
% reactive powers) per phase
subplot(2,2,1)
plot(l_3ph_profile(:,t-1:t+T_day-1)',':','linewidth',1)
hold on
ProfilAux = l_3ph_profile'...
       +[sum(xhat_a);sum(xhat_b);sum(xhat_c)]';
plot(ProfilAux(t-1:t+T_day-1,:),'linewidth',1)
hold off
grid
ylabel({'active power [kW]','(aggregated per phase)'})
xlabel('Forecast horizon [h]')
set(gca,'FontSize',12);
title( ['day = ' num2str(ceil((t*dt+h_0)/24)) '    hour = '...
   num2str(mod((t-2)*dt+h_0,24)) 'h'])

subplot(2,2,2)
plot(q_3ph_profile(:,t-1:t+T_day-1)',':','linewidth',1)
hold on
ProfilAux = q_3ph_profile'...
       +[sum(qhat_a);sum(qhat_b);sum(qhat_c)]';
plot(ProfilAux(t-1:t+T_day-1,:),'linewidth',1)
hold off
grid
ylabel({'reactive power [kVAr]','(aggregated per phase)'})
xlabel('Forecast horizon [h]')
set(gca,'FontSize',12);
title( ['Exchange = '...
   num2str(exchange) '     # of PEVs = ' num2str(length(index))])

% plot PEV load (active and reactive powers) per phase per PEV
subplot(2,2,3)
plot(xhat_a(:,t-1:t+T_day-1)','linewidth',1)
hold on
plot(xhat_b(:,t-1:t+T_day-1)','linewidth',1)
plot(xhat_c(:,t-1:t+T_day-1)','linewidth',1)
hold off
grid
ylabel({'active power [kW]','(per phase per PEV)'})
xlabel('Forecast horizon [h]')
set(gca,'FontSize',12);

subplot(2,2,4)
plot(qhat_a(:,t-1:t+T_day-1)','linewidth',1)
hold on
plot(qhat_b(:,t-1:t+T_day-1)','linewidth',1)
plot(qhat_c(:,t-1:t+T_day-1)','linewidth',1)
hold off
grid
ylabel({'reactive power [kVAr]','(per phase per PEV)'})
```

```
              xlabel('Forecast horizon [h]')
              set(gca,'FontSize',12);

              pause(1e-3)

          end
          % update the initial states of charge of the connected PEVs
          % for the next time step
          soc_0D(index) = soc_0D(index)...
                             +dt*(xhat_a(index,t)+xhat_b(index,t)+xhat_c(index,t));
      end
  end
```

5.9.4 Additional Script Using Logarithmic Barrier Functions

This final script is intended to be used for energy distributions with three-phase chargers. It can be extended to be used for single-phase chargers. The main difference with the scripts of Sect. 5.9.1 is that constraints (5.15a) and (5.15b) are not expressed with the escort functions. Instead these are expressed with logarithmic barrier functions included in the objective [BV04]. With this modification, the ED routine avoids boundaries that evolve with distributions. In fact the only boundaries represented by escort functions are those linked to nominal charging rates. We encourage the reader to check the differences in performance.

```
function xhat = Book_ED_P_3ph_barrier(soc_d,soc_0,soc_max,soc_min,p_max,l_max,dt,...
                          Ki,Lhat_,x0,mu,eta,init_steps)

% ------ Desciption inputs
% soc_d : final desired state of charge [kWh]
% soc_0 : initial state of charge [kWh]
% soc_max : max allowed state of charge in the connection time frame [kWh]
% soc_min : min allowed state of charge in the connection time frame [kWh]
% p_max : max charging/discharging rate [kW] (per phase)
% l_max : max transformed load allowed [kW]
% dt : time step
% Ki : total time steps
% Lhat_ : Load forecast from (t_0-1) to the available horizon (>Ki+1)
%          (array of 3 rows and Ki+2 columns or more)
% x0 : previously defined load distribution
%          (array of 3 rows and Ki columns)
% mu : owner's chosen trade-off factor in the interval [0,1]
% eta : grid operator chosen trade-off factor in the interval [0,1]
%          ("profile smoothing" or "load shifting - peak shaving")
% init_steps : (1) first round of agregator -PEV information  exhange
%              (2) second round of agregator -PEV information  exhange,
%                   useful for reseting initial distribution if
%                   transformed is constrained
%              (0) normal ED routine

% ------ Description outputs
% xhat : final load distribution for the three phases
%          (array of 3 rows and Ki columns)

% load profile (from aggregator) for the PEV connection time
Lhat = Lhat_(:,2:Ki+1);
```

```
% Different from MSD, in the ED routine Gamma and psk are energy
% not active power
Gamma = soc_d-soc_0;

% Threshold for accomodating load when the transformer has reached or will
% reach its limit at a given time slot
threshold = 0.1;

% Arrays of non-changing constraints
% (Sect. 5.6.1, Eqs. 5.15c and d)
p_up1 = dt*ones(3,Ki)*p_max; % nominal from charger
p_up2 = dt*max((l_max-(Lhat-x0)),threshold); % available from transformer

% variance tolerance
tol = 1e-3;

% Parameters of backtracking line search
alpha_ls = 0.5;
beta_ls = 0.1;

% Number of internal ED routine iterations
It = 1000;

% vector of upper and lower constraints
p_up = min(p_up1,p_up2); % charger and transformer
p_lo = -dt*ones(3,Ki)*p_max; % nominal from charger

% Definition of initial population distribution
% This is defined in two steps:
% 1) (init_steps=1) The distribution is initialized as a uniform
%    distribution over the connection time frame, and then it is sent to
%    the aggregator.
% 2) (init_steps=2) Once the aggregator receives all the initial
%    distributions, it returns the new aggregated total load profile. In
%    this second step, the PEV resets its load distribution according to
%    the availability of the transformer during the connection time frame.
if init_steps == 1
    % -- First initialization (first step)
    % distribution is defined using charger's nominal constraints (p_up1)
    psk_ = (p_up1/sum(sum(p_up1)))*(soc_d-soc_0);
    % the distribution is assigned
    psk = psk_;
    % after this, the distribution is sent to the aggregator without further
    % treatment
elseif init_steps == 2
    % -- Second initialization (second step) taking into account possible
    % transformer constraints
    if sum(sum(p_up)) > Gamma
        % If constraints from charger and transformer allow it, load is
        % distributed such that the initally required energy for the PEV is
        % consumed during the connection time frame
        psk_ = (p_up/sum(sum(p_up)))*(soc_d-soc_0);
    else
        % otherwise, constraints indicate that there is a lack of energy
        % available. Thus the PEV consumes what is available in a load
        % distribution given by the imposed constrains
        psk_ = (p_up/sum(sum(p_up)))*(sum(sum(p_up_))-1e-3);
        % thus the population size is smaller that the initially defined
        Gamma = sum(sum(p_up))-1e-3;
    end

    % the second reset distribution is assigned
```

```
        psk = psk_;
        % after this, the distribution is re-sent to the aggregator without
        % further treatment
    else
        % After the initial resets, load is again distributed according to
        % constraints as in step 2
        if sum(sum(p_up)) > Gamma
            % enough available energy
            psk_ = (p_up/sum(sum(p_up)))*(soc_d-soc_0);
        else
            % lack of energy available
            psk_ = (p_up/sum(sum(p_up)))*(sum(sum(p_up))-1e-3);
            % thus the population size is smaller that the initially defined
            Gamma = sum(sum(p_up))-1e-3;
        end
        % load distribution is assigned
        psk = psk_;
        % then, it is processed by the ED routine
    end

% definition of sigmas (Sect. 5.4)
sigma_up = Gamma-sum(sum(p_up));
sigma_lo = Gamma-sum(sum(p_lo));

% escort functions and sum of escort functions (Sect. 5.4)
Escort = (psk_-p_up).*(psk_-p_lo)/(sigma_up*sigma_lo);
SumEscort = sum(sum(Escort));

% matrix A_soc is useful to compute the state of charge from the
% distribution
Pre_A_soc = tril(ones(Ki));
Pre_A_soc(end,:) = [];
A_soc = repelem(Pre_A_soc,1,3);
% inverse weight parameter for barrier function
tt = 1000;


% computes euclidean gradient of logarithmic barrier function
fk_soc = A_soc'*(1./(soc_min-soc_0-A_soc*psk_(:))...
        +1./(soc_max-soc_0-A_soc*psk_(:)));
fk_soc = fk_soc([(1:Ki)*3-2; (1:Ki)*3-1 ;(1:Ki)*3])';

% payoff functions for pure strategies (Sect. 5.6.2)
bc = (Lhat_(:,1:Ki+2)-[zeros(3,1) x0 zeros(3,1)])';
pi_now = psk_'/dt+bc(2:Ki+1,:);
pi_pre = [zeros(3,1) psk_(:,1:Ki-1)]'/dt+bc(1:Ki,:);
pi_pos = [psk_(:,2:Ki) zeros(3,1)]'/dt+bc(3:Ki+2,:);

fk = (-2*eta*mu*(pi_now)-(1-eta)*mu*2*(-pi_pre+2*pi_now-pi_pos)...
    -2*(1-mu)*(psk_'/dt));

% adds euclidean gradient of logarithmic barrier function
fk= fk -(1/tt)*fk_soc;

% scalar field corresponding to payoff functions (based on Sect. 5.6.2)
F = (mu*(eta*sum(sum(pi_now.^2))+(1-eta)*(sum(sum((pi_now-pi_pre).^2,2)...
    +(pi_pos(:,end)-pi_now(:,end)).^2)))+(1-mu)*sum(sum((psk_'/dt).^2)));

% adds logarithmic barrier function to the original scalar field
F = F -(1/tt)*sum(log(-soc_min+soc_0...
                    +A_soc*psk_(:))+log(soc_max-soc_0-A_soc*psk_(:)));

% weighted average payoff
fmeank = trace(Escort*fk)/SumEscort;
```

```
if init_steps == 0
    % if the first two rounds of exchanges (verifying transformer's
    % constraints) have already passed, then the ED routine is run
    for i = 1:It

        % the new distribution is assigned
        psk = psk_;

        %% Subroutine for the backtracking line search
        Ts = 1e10;

        F_ir = Inf;
        % the descent direction (Delta_x) is defined by the ED gradient
        Delta_x = Escort.*(fk'-fmeank);
        if norm(Delta_x(:)) == 0
            % it only occurs when the equilibrium has been perfectly reached
            break
        else
            Delta_x = Delta_x/norm(Delta_x(:)); % normalize the ED gradient
        end
        if sum(sum(isnan(Delta_x)))>0
            break
        end

        while (F_ir > (F - alpha_ls*Ts*trace(fk'*Delta_x')))
            psk_ir = psk_ + Ts*Delta_x; % test distribution

            % test state of charge profile
            soc_k_ir = repmat(soc_0+cumsum(sum(psk_ir)),3,1);

            % checks if test distributions lie within specified ranges
            if (sum(sum(psk_ir<p_lo)) + sum(sum(psk_ir>p_up)))...
               + sum(soc_k_ir(1,:)<=soc_min)+ sum(soc_k_ir(1,:)>=soc_max))>0
                % it is useless to update the scalar field value if
                % parameters are out of range
            else
                % updates scalar field test value
                pi_now_ir = psk_ir'/dt + bc(2:Ki+1,:);
                pi_pre_ir = [zeros(3,1) psk_ir(:,1:Ki-1)]'/dt + bc(1:Ki,:);
                pi_pos_ir = [psk_ir(:,2:Ki) zeros(3,1)]'/dt + bc(3:Ki+2,:);
                F_ir =(mu*(eta*sum(sum(pi_now_ir.^2))...
                        +(1-eta)*(sum(sum((pi_now_ir-pi_pre_ir).^2,2)...
                        +(pi_pos_ir(:,end)-pi_now_ir(:,end)).^2)))...
                        +(1-mu)*sum(sum((psk_ir'/dt).^2)));
                % adds test log barrier function
                F_ir = F_ir-(1/tt)*sum(log(-soc_min+soc_0+A_soc*psk_ir(:))...
                                    +log(soc_max-soc_0-A_soc*psk_ir(:)));
            end
            % reduces step size Ts
            Ts = beta_ls*Ts;

            % If Ts is too small, the subroutine stops
            if Ts < 1e-30
                break
            end
        end
        if Ts < 1e- 30% if Ts is too small, the ED routine is stopped
            break
        end

        %%
```

```matlab
        % Computes next distribution
        psk_ = psk_+Ts*Delta_x;

        % computes next escort function values and their sum
        Escort = (psk_-p_up).*(psk_-p_lo)/(sigma_up*sigma_lo);
        SumEscort = sum(sum(Escort));

        % computes euclidean gradient of logarithmic barrier function
        fk_soc = A_soc'*(1./(soc_min-soc_0-A_soc*psk_(:))...
                        +1./(soc_max-soc_0-A_soc*psk_(:)));
        fk_soc = fk_soc([(1:Ki)*3-2; (1:Ki)*3-1 ;(1:Ki)*3])';

        % new payoffs
        pi_now = psk_'/dt+bc(2:Ki+1,:);
        pi_pre = [zeros(3,1) psk_(:,1:Ki-1)]'/dt+bc(1:Ki,:);
        pi_pos = [psk_(:,2:Ki) zeros(3,1)]'/dt+bc(3:Ki+2,:);

        fk = (-2*eta*mu*(pi_now)-(1-eta)*mu*2*(-pi_pre+2*pi_now-pi_pos)...
             -2*(1-mu)*(psk_'/dt));

        % adds euclidean gradient of logarithmic barrier function
        fk = fk -(1/tt)*fk_soc;

        % new scalar field value corresponding to payoff functions
        F = (mu*(eta*sum(sum(pi_now.^2)))...
            +(1-eta)*(sum(sum((pi_now-pi_pre).^2,2)...
            +(pi_pos(:,end)-pi_now(:,end)).^2)))...
            +(1-mu)*sum(sum((psk_'/dt).^2)));

        % adds logarithmic barrier function to the original scalar field
        F = F -(1/tt)*sum(log(-soc_min+soc_0...
                +A_soc*psk_(:))+log(soc_max-soc_0-A_soc*psk_(:)));

        % weighted average payoff
        fmeank = trace(Escort*fk)/SumEscort;

        % updates the weighted variance stopping criteria
        variancef = sum(Escort(:).*(fk(:)...
                        -sum(Escort(:).*fk(:))/SumEscort).^2)/SumEscort;

        % stopping criteria: variance, deviations, or max iterations
        if (SumEscort*variancef < tol) || ((sum(sum(psk_))/Gamma)-1)^2>0.01
            tt = 10*tt;
        end

        % If you want to check the evolution inside this ED routine
        % please uncomment this sectionof code
%         figure(10)
%         plot(bc,':')
%         hold on
%         plot(bc+[zeros(3,1) psk/dt zeros(3,1)]')
%         hold off
%         grid
%         ylabel('active power [kW]')
%         xlabel('forecast horizon [h]')
%         set(gca,'FontSize',12);
%         title( ['iteration = ' num2str(i) '    tt = '...
%                num2str(tt)])
%         pause(1e-6)

    end
end
% output assignation
```

```
xhat = psk/dt;
end
```

5.10 Conclusion

Escort evolutionary game dynamics is proposed as a tool for the distributed optimization application to the integral load management of PEVs. Several details and features of the ED are presented, as well as the proposed use of escort functions. The employed analogies, and specially the proposed ideas behind the treatment given to the intersection of both upper and lower constraints simplices, can be adapted to problems of different nature.

Several illustrative examples are introduced to clarify the differences between the ED and the MSD approach of Chap. 4. Even if these examples are useful to observe that ED exhibits a less computationally expensive performance than MSD, the real disadvantage of MSD lies in the problem of finding all the vertices of the intersection, specially when the number of dimensions is increased, and when the constraints are not homogeneous for all the variables.

The final examples are useful to highlight the fact that the number of vertices, defining the intersection set, can be unpredictable specially in higher dimensions, with heterogeneous constraints, and in cases where the simplices change with time as it is the case for the proposed PEV load management application of this chapter. Nevertheless, for applications like the one proposed in Sect. 4.5.2 of the previous chapter, where the most convenient mixed strategies (for PEV owners convenience) are employed, it is not obvious to apply a strategy like the one proposed in this chapter with ED. It is necessary to highlight that both approaches have their own advantages and benefits for the distributed PEV load management problem.

The next chapter takes advantage of the features of the proposed approach to propose an integral PEV load management methodology. While batteries of PEVs are guaranteed to be fully recharged, the proposed scheme is such that PEVs also work together in a fair scheme, to balance the active power load, and partially supply the reactive power demand of each phase of the distribution system transformer. Again, it is important to mention that the ED with the proposed escort functions may have applications on other domains where distributed optimization is required.

The proposed approach decentralizes the optimization procedures and coordinates the replies of each PEV, and their associated fictive populations, such that the evolution provides optimal schedules for each PEV while ancillary services are provided. The PEV load scheduling problem is modeled as a multi-population scenario where each population evolve and redistribute their individuals in several environments according to the benefits they provide. The final part of this chapter was devoted to evaluate the proposed approach under several scenarios, and test how it works and the benefits it provides to both the PEV owners and the grid.

Assuming different interests of the utility grid manager (reflected on values of parameter η), and the efficiency of its policies to motivate PEV owners (reflected on the average value of parameters μ^i), the methodology is tested with the presence of only three-phase chargers. It was found that a high presence of PEVs becomes very useful for both the grid and the PEV owners. During periods with high penetration of PEVs, services of reactive power compensation, and active and reactive power balancing among phases, are provided more efficiently than it is observed on hours of low presence.

The methodology is also tested under presence of only single-phase chargers. With only three-phase chargers, tasks are likely to be fairly shared by PEVs given that all of them have the same objectives and their actions similarly affect other PEVs. With single-phase chargers, actions of a PEV affect only those PEVs connected to the same phase. Then the concept of fairness is managed differently, with the second utility grid parameter v. In one side, if $v = 0$, ancillary services are provided taking into account only the total active and reactive power profiles, and neglecting the effects on each phase separately. A service trade-off parameter fixed at $v = 0$ is useful to get a fair decentralized allocation of charging rates, among all the connected PEVs without interest on the phase of connection. Meanwhile the smoothing/flattening services are provided to the total active and reactive power profiles.

With a low penetration of PEVs and $v = 1$, PEVs will be in charge of compensating variations on their corresponding phases. These variations are less strong than they could be for total active and reactive power profiles. Besides, it can be noticed that the smoothing/flattening objectives for individual phases, provide good results for total profiles at the same time.

Example 4 presents a combination of both single-phase and three-phase chargers, which is more likely to occur. In such a case, it was observed that the presence of three-phase chargers mitigates the effects that the service trade-off parameter v has on the behavior of PEVs connected to single-phase chargers. The analysis of the considered scenarios shows that the the smoothing, flattening and balancing objectives are achieved almost identically when $v = 0$ and when $v = 1$ if there is presence of both single and three-phase chargers. However, v has a *clusterization* effect on the concept of fairness. When $v = 0$, charging rates are fairly allocated among PEVs, regardless of their type of charger. On the other hand, when $v = 1$ charging rates are still fairly allocated, but in this case fairness is observed per groups depending on their type of charger and the phase where they are connected (4 groups in total).

For a deeper analysis of the sensitivity of the methodology under variation of the control parameters (η, μ, v), we encourage readers to modify and test the scripts provided in Sect. 5.9, at the end of this chapter.

References

[EP15]   Electric Power Research institute (EPRI). *Plug-in Electric Vehicle Projections: Scenarios and Impacts*. Tech. rep. 2015 (cit. on p. 3).

[AF96]   David Avis and Komei Fukuda. "Reverse search for enumeration". In: *Discrete Applied Mathematics* 65.1-3 (1996). First International Colloquium on Graphs and Optimization, pp. 21–46 (cit. on p. 144).

[Age16]   International Energy Agency. *Global EV Outlook 2016, Beyond One million electric cars*. Tech. rep. 2016 (cit. on p. 3).

[Avi14]   Avicenne. *Battery Market Development for Consumer Electronics, Automotive, and Industrial*. Tech. rep. 2014 (cit. on p. 3).

[AWA07]   H. Akagi, E.H. Watanabe, and M. Aredes. *Instantaneous Power Theory and Applications to Power Conditioning*. IEEE Press Series on Power Engineering. Wiley, 2007 (cit. on p. 14).

[Bas+15]   K. Basu, V. Debusschere, S. Bacha, U. Maulik, and S. Bondyopadhyay. "NonintrusiveLoad Monitoring: A Temporal Multilabel Classification Approach". In: *IEEE Trans. Ind. Informat.* 11.1 (Feb. 2015), pp. 262–270 (cit. on pp. 3, 79, 158).

[Ber12]   D.P. Bertsekas. *Dynamic Programming and Optimal Control: Approximate dynamic programming*. Athena Scientific optimization and computation series. Athena Scientific, 2012 (cit. on pp. 69, 71).

[Bil+14]   E. Bilbao, P. Barrade, I. Etxeberria-Otadui, A. Rufer, S. Luri, and I. Gil. "Optimal Energy Management Strategy of an Improved Elevator With Energy Storage Capacity Based on Dynamic Programming". In: *IEEE Transactions on Industry Applications* 50.2 (Mar. 2014), pp. 1233–1244 (cit. on p. 69).

[Bou+17]   Anouar Bouallaga, Arnaud Davigny, Vincent Courtecuisse, and Benoit Robyns. "Methodology for technical and economic assessment of electric vehicles integration in distribution grid". In: *Mathematics and Computers in Simulation* 131 (2017). 11th International Conference on Modeling and Simulation of Electric Machines, Converters and Systems, pp. 172–189 (cit. on p. 6).

[BV00]   A.R. Bergen and V. Vittal. *Power Systems Analysis*. Pearson/Prentice Hall, 2000 (cit. on pp. 14, 15, 87). 2000

[BV04]   S.P. Boyd and L. Vandenberghe. *Convex Optimization*. Cambridge University Press, 2004 (cit. on pp. 41, 44, 107, 122, 125, 159, 204).

[CC15]   United Nations Framework Convention on Climate Change. *Adoption of the Paris agreement*. Tech. rep. 2015 (cit. on p. 3).

[CNHD10]   K. Clement-Nyns, E. Haesen, and J. Driesen. "The Impact of Charging Plug-In Hybrid Electric Vehicles on a Residential Distribution Grid". In: *IEEE Trans. Power Syst.* 25.1 (Feb. 2010), pp. 371–380 (cit. on pp. 4, 69).

© Springer International Publishing AG 2018
A. Ovalle et al., *Grid Optimal Integration of Electric Vehicles: Examples with Matlab Implementation*, Studies in Systems, Decision and Control 137,
https://doi.org/10.1007/978-3-319-73177-3

[CNHD11] K. Clement-Nyns, E. Haesen, and J. Driesen. "The impact of vehicle-to-grid on the distribution grid". In: *Electric Power Systems Research* 81.1 (2011), pp. 185–192 (cit. on pp. 4, 69).

[D'E05] Jhon D'Errico. *Permutations of repeated elements*. https://fr.mathworks.com/ matlabcentral/newsreader/view_thread/102182. Aug. 2005 (cit. on p. 123).

[DLP08] Davide Dragone, Luca Lambertini, and Arsen Palestini. "A Class of Best-Response Potential Games". In: *Working Paper DSE 635, Department of Economics, University of Bologna* (June 2008) (cit. on p. 78).

[DSB10] K.J. Dyke, N. Schofield, and M. Barnes. "The Impact of Transport Electrification on Electrical Networks". In: *IEEE Trans. Ind. Electron.* 57.12 (Dec. 2010), pp. 3917–3926 (cit. on p. 3).

[EUR14] EUROBAT. *A review of battery technologies for automotive applications*. Tech.rep. 2014 (cit. on p. 3).

[FHB15] J. A. Fernandez, A. Hably, and A. I. Bratcu. "Assessing the economic profit of a vehicle-to-grid strategy for current unbalance minimization". In: *2015 IEEE International Conference on Industrial Technology (ICIT)*. Mar. 2015, pp. 2628–2635 (cit. on p. 7).

[GM15] D. Gregoratti and J. Matamoros. "Distributed Energy Trading: The Multiple-Microgrid Case". In: *IEEE Trans. Ind. Electron.* 62.4 (Apr. 2015), pp. 2551–2559 (cit. on p. 94).

[GTL13] Lingwen Gan, U. Topcu, and S.H. Low. "Optimal decentralized protocol for electric vehicle charging". In: *IEEE Trans. Power Syst.* 28.2 (May 2013), pp. 940–951 (cit. on p. 7).

[Gue+05] J.M. Guerrero, L. Garcia de Vicuna, J. Matas, M. Castilla, and J. Miret. "Output Impedance Design of Parallel-Connected UPS Inverters With Wireless Load Sharing Control". In: *IEEE Trans. Ind. Electron.* 52.4 (2005), pp. 1126–1135 (cit. on pp. 15, 87).

[Hae+14] Pierre Haessig, Thibaut Kovaltchouk, Bernard Multon, Hamid Ben Ahmed, and Stéphane Lascaud. "Computing an Optimal Control Policy for an Energy Storage". In: *CoRR* abs/1404.6389 (2014) (cit. on p. 69).

[Har09] Marc Harper. "Information Geometry and Evolutionary Game Theory". In: *CoRR* abs/0911.1383 (2009) (cit. on pp. 99, 100, 107, 143, 148).

[Har11] Marc Harper. "Escort evolutionary game theory". In: *Physica D: Nonlinear Phenomena* 240.18 (2011), pp. 1411–1415 (cit. on pp. 136, 140, 143, 147, 148, 159).

[HS88] J. Hofbauer and K. Sigmund. *The Theory of Evolution and Dynamical systems*. Cambridge University Press, 1988 (cit. on pp. 78, 93, 94, 98, 99, 107, 141, 143, 158).

[HS98] J. Hofbauer and K. Sigmund. *Evolutionary Games and Population Dynamics*. Cambridge University Press, 1998 (cit. on pp. 141, 143).

[HVG12] Yifeng He, B. Venkatesh, and Ling Guan. "Optimal Scheduling for Charging and Discharging of Electric Vehicles". In: *IEEE Trans. Smart Grid* 3.3 (Sept. 2012), pp. 1095–1105 (cit. on p. 6).

[Iee] *IEEE PES Distribution Systems Analysis Subcommittee, Radial Test Feeders*. URL: https://ewh.ieee.org/soc/pes/dsacom/testfeeders/ (cit. on pp. 11, 25–27, 81).

[JTG13] Chenrui Jin, Jian Tang, and P. Ghosh. "Optimizing Electric Vehicle Charging: A Customer's Perspective". In: *IEEE Trans. Veh. Technol.* 62.7 (Sept. 2013), pp. 2919–2927 (cit. on pp. 4, 94).

[Ker01] W.H. Kersting. *Distribution System Modeling and Analysis*. The electric power engineering series. CRC Press, 2001 (cit. on p. 14).

[Kin92] J.F.C. Kingman. *Poisson Processes*. Oxford Studies in Probability. Clarendon Press, 1992 (cit. on p. 118).

[KK89] H. Kesavan and J.N. Kapur. "The generalized maximum entropy principle". In: *IEEE Trans. Syst., Man and Cybernet.* 19.5 (Sept. 1989), pp. 1042–1052 (cit. on pp. 94, 104, 105).

[Kra+16] I. Krastev, P. Tricoli, S. Hillmansen, and M. Chen. "Future of Electric Railways: Advanced Electrification Systems with Static Converters for ac Railways". In: *IEEE Electrification Magazine* 4.3 (Sept. 2016), pp. 6–14 (cit. on p. 3).

[LD08]   M. Lara and L. Doyen. *Sustainable Management of Natural Resources: Mathematical Models and Methods*. Environmental Science and Engineering. Springer, 2008 (cit. on p. 69).

[Leb06]   G. Lebanon. "Metric learning for text documents". In: *IEEE Trans. Pattern Anal. Mach. Intell.* 28.4 (Apr. 2006), pp. 497–508 (cit. on pp. 99, 100, 140).

[Liu+15]   Yi Liu, Chau Yuen, N. Ul Hassan, Shisheng Huang, Rong Yu, and Shengli Xie. "Electricity Cost Minimization for a Microgrid With Distributed Energy Resource Under Different Information Availability". In: *IEEE Trans. Ind. Electron.* 62.4 (Apr. 2015), pp. 2571–2583 (cit. on p. 94).

[LLL14]   J. Lin, K. C. Leung, and V. O. K. Li. "Optimal Scheduling With Vehicle-to-Grid Regulation Service". In: *IEEE Internet Things J.* 1.6 (Dec. 2014), pp. 556–569 (cit. on p. 8).

[LSH10]   M. Liserre, T. Sauter, and J. Y. Hung. "Future Energy Systems: Integrating Renewable Energy Sources into the Smart Power Grid Through Industrial Electronics". In: *IEEE Industrial Electronics Magazine* 4.1 (Mar. 2010), pp. 18–37 (cit. on p. 3).

[MCH13]   Zhongjing Ma, D.S. Callaway, and I.A. Hiskens. "Decentralized Charging Control of Large Populations of Plug-in Electric Vehicles". In: *IEEE Trans. Control Syst. Technol.* 21.1 (Jan. 2013), pp. 67–78 (cit. on p. 7).

[Mei+13]   P. Meibom, K. B. Hilger, H. Madsen, and D. Vinther. "Energy Comes Together in Denmark: The Key to a Future Fossil-Free Danish Power System". In: *IEEE Power and Energy Magazine* 11.5 (Sept. 2013), pp. 46–55 (cit. on p. 3).

[Men+95]   Anil Menon, Kishan Mehrotra, Chilukuri K. Mohan, and Sanjay Ranka. "Optimization Using Replicators". In: *Proceedings of the 6th International Conference on Genetic Algorithms, Pittsburgh, PA, USA, July 15-19, 1995*. 1995, pp. 209–216 (cit. on p. 141).

[Men01]   Anil Menon. "Replicators & Complementarity: Solving the Simplest Complex System without Simulation." In: *International Conference Computational Science ICCS 2001, San Francisco, CA, USA, May 28-30, 2001. Proceedings, Part II*. Vol. 2074. 2001, pp. 922–934 (cit. on pp. 94, 99).

[MM15]   Davis Montenegro Martinez. "Actor's based diakoptics for the simulation, monitoring and control of smart grids". Theses. Université Grenoble Alpes, Nov. 2015(cit. on p. 3).

[Ngu+14a] V. L. Nguyen, T. Tran-Quoc, S. Bacha, and N. A. Luu. "Charging strategies to minimize the energy cost for an electric vehicle fleet". In: *IEEE PES Innovative Smart Grid Technologies, Europe*. Oct. 2014, pp. 1–7 (cit. on p. 5).

[Ngu+14b] V. L. Nguyen, T. Tran-Quoc, S. Bacha, and B. Nguyen. "Charging strategies to minimize the peak load for an electric vehicle fleet". In: *IECON 2014 - 40th Annual Conference of the IEEE Industrial Electronics Society*. Oct. 2014, pp. 3522–3528 (cit. on p. 5).

[NS12]   Hung Khanh Nguyen and Ju Bin Song. "Optimal charging and discharging for multiple PHEVs with demand side management in vehicle-to-building". In: *Journal Communications and Networks* 14.6 (Dec. 2012), pp. 662–671 (cit. on pp. 7, 78, 79).

[OHB15]   A. Ovalle, A. Hably, and S. Bacha. "Optimal management and integration of electric vehicles to the grid: Dynamic programming and game theory approach". In: *2015 IEEE International Conference on Industrial Technology (ICIT)*. Mar. 2015, pp. 2673–2679 (cit. on pp. 67, 70, 72, 80, 81).

[Ope]   *Electric Power Research Institute, OpenDSS, Distribution System Simulator*. URL: https://sourceforge.net/projects/electricdss/ (cit. on pp. 25–27, 81).

[Ota+12]   Y. Ota, H. Taniguchi, T. Nakajima, K. M. Liyanage, J. Baba, and A. Yokoyama. "Autonomous Distributed V2G (Vehicle-to-Grid) Satisfying Scheduled Charging". In: *IEEE Trans. on Smart Grid* 3.1 (Mar. 2012), pp. 559–564 (cit. on p. 8).

[Ova+14]   A. Ovalle, A. Hably, S. Bacha, and M. Ahmed. "Voltage Support by Optimal Integration of Plug-in Hybrid Electric Vehicles to a Residential Grid". In: *IECON2014 - 40th Annual Conference on IEEE Industrial Electronics Society*. Nov. 2014 (cit. on pp. 11, 14, 21, 69).

[Ova+15]   A. Ovalle, G. Ramos, S. Bacha, A. Hably, and A. Rumeau. "Decentralized Control of Voltage Source Converters in Microgrids Based on the Application of Instantaneous Power Theory". In: *IEEE Trans. Ind. Electron.* 62.2 (Feb. 2015), pp. 1152–1162 (cit. on pp. 14, 15, 87, 94).

[Ova+16a]   A. Ovalle, J. Fernandez, A. Hably, and S. Bacha. "An Electric Vehicle Load Management Application of the Mixed Strategist Dynamics and the Maximum Entropy Principle". In: *IEEE Trans. Ind. Electron.* 63.5 (May 2016), pp. 3060–3071 (cit. on pp. 94, 106, 110–115, 144).

[Ova+16b]   A. Ovalle, S. Bacha, A. Hably, and K. Basu. "On the most convenient mixed strategies in a mixed strategist dynamics approach for load management of electric vehicle fleets". In: *IECON 2016 - 42nd Annual Conference of the IEEE Industrial Electronics Society.* Oct. 2016, pp. 2082–2088 (cit. on pp. 94, 101, 103, 116–122).

[Ova+17]   A. Ovalle, S. Bacha, A. Hably, G. Ramos, and J. M. Hossain. "Escort Evolutionary Game Dynamics Approach to Integral Load Management of Electric Vehicle Fleets". In: *IEEE Transactions on Industrial Electronics* 64.2 (Feb. 2017), pp. 1358–1369 (cit. on pp. 136, 146, 157, 159, 161).

[PDK12]   C. Pang, P. Dutta, and M. Kezunovic. "BEVs/PHEVs as Dispersed Energy Storage for V2B Uses in the Smart Grid". In: *IEEE Transactions on Smart Grid* 3.1 (Mar. 2012), pp. 473–482 (cit. on p. 8).

[Pel09]   Marcello Pelillo. "Replicator Dynamics in Combinatorial Optimization." In: *Encyclopedia of Optimization.* Ed. by Christodoulos A. Floudas and Panos M. Pardalos. Springer, 2009, pp. 3279–3291 (cit. on pp. 93, 94).

[PQ11]   A. Pantoja and N. Quijano. "A Population Dynamics Approach for the Dispatch of Distributed Generators". In: *IEEE Trans. Ind. Electron.* 58.10 (Oct. 2011), pp. 4559–4567 (cit. on p. 94).

[PQP14]   A. Pantoja, N. Quijano, and K.M. Passino. "Dispatch of distributed generators under local-information constraints". In: *American Control Conference (ACC), 2014.* June 2014, pp. 2682–2687 (cit. on p. 94).

[PS13]   E. Paparoditis and T. Sapatinas. "Short-Term Load Forecasting: The Similar Shape Functional Time-Series Predictor". In: *IEEE Trans. Power Syst.* 28.4 (Nov. 2013), pp. 3818–3825 (cit. on pp. 3, 79, 158).

[RFH14]   P. Rezaei, J. Frolik, and P.D.H. Hines. "Packetized Plug-In Electric Vehicle Charge Management". In: *IEEE Trans. Smart Grid* 5.2 (Mar. 2014), pp. 642–650 (cit. on p. 6).

[RFK12]   P. Richardson, D. Flynn, and A. Keane. "Local Versus Centralized Charging Strategies for Electric Vehicles in Low Voltage Distribution Systems". In: *IEEE Trans. Smart Grid* 3.2 (June 2012), pp. 1020–1028 (cit. on pp. 6, 7).

[Rif+11]   Y. Riffonneau, S. Bacha, F. Barruel, and S. Ploix. "Optimal Power Flow Management for Grid Connected PV Systems With Batteries". In: *IEEE Transactions on Sustainable Energy* 2.3 (July 2011), pp. 309–320 (cit. on pp. 69, 70).

[Ros65]   J. B. Rosen. "Existence and Uniqueness of Equilibrium Points for Concave N Person Games". In: *Econometrica* 33.3 (1965), pp. 520–534 (cit. on pp. 7, 78).

[Sar+16a]   S. Sarabi, A. Davigny, V. Courtecuisse, Y. Riffonneau, and B. Robyns. "Potential of vehicle-to-grid ancillary services considering the uncertainties in plug-in electric vehicle availability and service/localization limitations in distribution grids". In: *Applied Energy* 171 (2016), pp. 523–540 (cit. on p. 8).

[Sar+16b]   S. Sarabi, A. Davigny, Y. Riffonneau, and B. Robyns. "V2G electric vehicle charging scheduling for railway station parking lots based on binary linear programming". In: *2016 IEEE International Energy Conference (ENERGYCON).* Apr. 2016, pp. 1–6 (cit. on p. 5).

[SB10a]   O. Sundström and C. Binding. "Optimization methods to plan the charging of electric vehicle fleets". In: *Proc. Int. Conf. Control, Commun.,* Power Eng. 2010, 28–29 (cit. on p. 5).

[SB10b]   O. Sundstrom and C. Binding. "Planning electric-drive vehicle charging under con-
           strained grid conditions". In: *Power System Technology (POWERCON), 2010 Interna-
           tional Conference on*. Oct. 2010, pp. 1–6 (cit. on p. 5).
[SB12]   O. Sundstrom and C. Binding. "Flexible Charging Optimization for Electric Vehicles
           Considering Distribution Grid Constraints". In: *IEEE Transactions on Smart Grid* 3.1
           (Mar. 2012), pp. 26–37 (cit. on p. 5).
[SDL08]   William H. Sandholm, Emin Dokumacı, and Ratul Lahkar. "The projection dynamic
           and the replicator dynamic". In: *Games and Economic Behavior* 64.2 (2008). Special
           Issue in Honor of Michael B. Maschler, pp. 666–683 (cit. on p. 148).
[SES12]   E. Sortomme and M. A. El-Sharkawi. "Optimal Scheduling of Vehicle-to-Grid Energy
           and Ancillary Services". In: *IEEE Trans. on Smart Grid* 3.1 (Mar. 2012), pp. 351–359
           (cit. on p. 8).
[SOR]   SOREA. *Énergies et Communications, Electricity distribution company, Régionde
           Savoie, France*. http://www.sorea-maurienne.fr/ (cit. on pp. 109–111, 114, 116, 118,
           119, 134, 171, 173).
[TBH14]   H. Turker, S. Bacha, and A. Hably. "Rule-Based Charging of Plug-in Electric Ve-
           hicles (PEVs): Impacts on the Aging Rate of Low-Voltage Transformers". In: *IEEE
           Transactions on Power Delivery* 29.3 (June 2014), pp. 1012–1019 (cit. on pp. 3, 7).
[Tur+12]   H. Turker, S. Bacha, D. Chatroux, and A. Hably. "Low-Voltage Transformer Loss-of-
           Life Assessments for a High Penetration of Plug-In Hybrid Electric Vehicles (PHEVs)".
           In: *IEEE Transactions on Power Delivery* 27.3 (July 2012), pp. 1323–1331 (cit. on p.
           7).
[Voo00]   Mark Voorneveld. "Best-response potential games". In: *Economics Letters* 66.3(2000),
           pp. 289–295 (cit. on p. 78).
[Web07]   J.N. Webb. *Game Theory: Decisions, Interaction and Evolution*. Springer Undergrad-
           uate Mathematics Series. Springer, 2007 (cit. on p. 78).
[WOB14]   Y. Wang, B. O'Donoghue, and S. Boyd. "Approximate dynamic programming via iter-
           ated Bellman inequalities". In: *International Journal of Robust and Nonlinear Control*
           (2014) (cit. on p. 67).
[Wu+11]   Chenye Wu, H. Mohsenian-Rad, Jianwei Huang, and A.Y. Wang. "Demand side man-
           agement for Wind Power Integration in microgrid using dynamic potential game theo-
           ry". In: *2011 IEEE GLOBECOM Workshops (GC Wkshps)*. Dec. 2011, pp. 1199–1204
           (cit. on p. 78).
[XL15]   Yinliang Xu and Zhicheng Li. "Distributed Optimal Resource Management Based on
           the Consensus Algorithm in a Microgrid". In: *IEEE Trans. Ind. Electron.* 62.4 (Apr.
           2015), pp. 2584–2592 (cit. on p. 94).
[YK13]   M. Yilmaz and P. T. Krein. "Review of the Impact of Vehicle-to-Grid Technologies on
           Distribution Systems and Utility Interfaces". In: *IEEE Trans. Power Electron.* 28.12
           (Dec. 2013), pp. 5673–5689 (cit. on p. 8).
[Zha+12]   P. Zhang, K. Qian, C. Zhou, B. G. Stewart, and D. M. Hepburn. "A Methodology for
           Optimization of Power Systems Demand Due to Electric Vehicle Charging Load". In:
           IEEE Transactions on Power Systems 27.3 (Aug. 2012), pp. 1628–1636 (cit. on p. 6).

Printed in the United States
By Bookmasters